Introduction to Kinesiology

INTRODUCTION TO KINESIOLOGY

Edited by:

Ravi Pandya

Delve
PUBLISHING

www.delvepublishing.com

Introduction to Kinesiology

Ravi Pandya

Delve Publishing
224 Shoreacres Road
Burlington, ON L7L 2H2
Canada
www.delvepublishing.com
Email: orders@arclereducation.com

ISBN: 978-1-77469-508-1 (Hardcover)

ABOUT THE EDITOR

Dr. Ravi Pandya has completed his graduation in 2018 from Gujarat Ayurved University Jamnagar in 2014 & Post Graduation from Parul University in 2018. He has academic experience of 2.9 years and current working in Parul Institute of Ayurved, Parul University, Vadodara, Gujarat, as deputy medical superintendent and assistant professor in Department of Kriyasharir.

TABLE OF CONTENTS

List of Figures .... *xi*
List of Abbreviations .... *xv*
Abstract .... *xvii*
Preface .... *xix*

Chapter 1 Fundamentals of Kinesiology 1
1.1. Introduction 2
1.2. History of Kinesiology 4
1.3. Descriptive Terminology 9
1.4. Segments of the Body 12
1.5. Types of Motion 13
1.6. Joint Movements (Osteokinematics) 16
1.7. Kinesiology of Testing Muscles 22
1.8. Development of Traditional Kinesiology 24
References 26

Chapter 2 Kinesiology of Shoulder Complex 33
2.1. Introduction 34
2.2. Structure of the Bones of the Shoulder Complex 34
2.3. The Shoulder Complex's Joint Structure and Supporting Structures 47
2.4. Muscle Activity in the Shoulder Complex: Pathomechanics and Mechanics 57
2.5. The Relationship Between Joint and Muscle Force Analysis and Clinical Practice 66
References 67

Chapter 3 Kinesiology of Elbow Units 75
3.1. Introduction 76
3.2. Structure of the Bones of Elbow 76

3.3. Articulations and Supporting Structures of the Elbow 84
3.4. Mechanics and Pathomechanics of Muscle Activity at the Elbow 87
3.5. Comparisons Among the Elbow Flexors 91
References 97

Chapter 4 **Kinesiology of Hand 105**
4.1. Introduction 106
4.2. Joints and Motions of the Thumb 106
4.3. Joints and Motions of the Fingers 107
4.4. Ligaments and Other Structures 109
4.5. Muscles of the Thumb and Fingers 111
References 117

Chapter 5 **Kinesiology of Gait 121**
5.1. Introduction 122
5.2. Analysis of Stance Phase 126
5.3. Analysis of Swing Phase 128
5.4. Additional Determinants of Gait 130
5.5. Age-Related Gait Patterns 132
5.6. Abnormal (Atypical) Gait 132
References 141

Chapter 6 **Kinesiology of Neck and Trunk 149**
6.1. Introduction 150
6.2. Joint Motions 151
6.3. Bones and Landmarks 152
6.4. Joints and Ligaments 156
6.5. Muscles of the Neck and Trunk 158
6.6. Muscles of the Cervical Spine 159
6.7. Muscles of the Trunk 161
6.8. Anatomical Relationships 164
References 166

Chapter 7 A Kinesiology of Hip Joint.......171
7.1. Introduction.......172
7.2. Joint Motions and Structure.......172
7.3. Landmarks and Bones.......174
7.4. Ligaments and Other Structures.......177
7.5. Muscles of the Hip.......181
7.6. Anatomical Relationships.......184
7.7. Common Hip Pathologies.......185
7.8. An Overview of Muscle Innervation.......187
References.......188

Chapter 8 Kinesiology of Ankle Joint and Foot.......193
8.1. Introduction.......194
8.2. Functional Characteristics of the Foot.......197
8.3. Joints and their Motions.......198
8.4. Ligaments and Other Structures.......204
8.5. Muscles of the Foot and Ankle.......206
8.6. Anatomical Associations.......210
8.7. Summary of Muscle Innervation.......212
8.8. Usual Ankle Pathologies.......212
References.......215

Index.......223

LIST OF FIGURES

Figure 1.1. Descriptive positions

Figure 1.2. The triad of health

Figure 1.3. Descriptive terminology

Figure 1.4. A quadruped's descriptive terminology

Figure 1.5. Body segments

Figure 1.6. Curvilinear motion

Figure 1.7. Rectilinear motion

Figure 1.8. Angular motion

Figure 1.9. Motion that combines angular and linear motion

Figure 1.10. Flexion and extension of the joints

Figure 1.11. Joint motions of abduction and adduction

Figure 1.12. Circumduction motion

Figure 1.13. Joint rotation motions

Figure 1.14. Inversion and eversion of left foot

Figure 1.15. Protraction and retraction

Figure 2.1. Clavicle. (A) View of the superior surface; (B) view of the inferior surface

Figure 2.2. Location of the scapula. (A) The scapula has been more posteriorly placed in humans. (B) In quadrupedal animals, the scapula is positioned on the lateral portion of the thorax

Figure 2.3. Scapula. (A) Anterior surface; (B) posterior surface

Figure 2.4. Plane of the scapula. The plane of the scapula makes an angle of roughly 40° with the frontal plane when seen transversely

Figure 2.5. The scapula rotates. The glenoid fossa faces upward (2) or downward (3) depending on whether the scapula is rotated about an anterior-posterior (AP) axis (3)

Figure 2.6. Scapular rotation. The scapula tilts anteriorly and posteriorly when rotated about an ML axis

Figure 2.7. Proximal humerus. (A) Anterior view; (B) posterior view

Figure 2.8. Orientation of the humerus's head. (A) The humeral head is rotated posteriorly in the transverse plane with the distal humeral condyles. (B) In the frontal plane, the humeral head is inclined superiorly and medially about the humeral shaft

Figure 2.9. This is the articular surface of the sternum. The sternum supplies the clavicle's head with a shallow articular surface

Figure 2.10. The shape of the thorax. The scapula's mobility is influenced by the elliptical form of the thorax

Figure 2.11. The sternoclavicular joint. The intraarticular disc, the capsule, the interclavicular ligament, the posterior and anterior sternoclavicular ligaments, and the costoclavicular ligament are the supporting components of the sternoclavicular joint

Figure 2.12. Forces that tend to medially shift the clavicle. A fall on the lateral side of the shoulder creates a force that tends to press the clavicle medially

Figure 2.13. A common method of clavicle fracture

Figure 2.14. The sternoclavicular joint's axes of movement

Figure 2.15. With one another, the articular surfaces of the acromioclavicular joint are comparatively beveled and flattened

Figure 2.16. The shoulder complex's superficial muscles

Figure 2.17. Trapezius muscle. The trapezius is split into three sections: lower, middle, and upper

Figure 3.1. The distal humerus, proximal ulna, and proximal radius make up the elbow joint complexity

Figure 3.2. How well the distal humerus flattens anteriorly and posteriorly in a trans image of the midshaft of the humerus and the distal humerus

Figure 3.3. The forward curve of the humerus' distal end in a sagittal perspective

Figure 3.4. The articular surface of the trochlear groove

Figure 3.5. An anterior view of the proximal radius

Figure 3.6. The radius in a bow form

Figure 3.7. Humeroulnar articular surfaces

Figure 3.8. The biceps brachii, brachialis, brachioradialis, and pronator teres are the major flexion muscles of the forearm

Figure 3.9. The widths of the brachioradialis, biceps brachii, and brachialis muscles in contrast

Figure 4.1. Bones and joints of the thumb and fingers

Figure 4.2. The carpometacarpal (CMC) joints of the fingers and thumb

Figure 4.3. Transverse and palmar carpal ligaments constitute the flexor retinaculum

Figure 4.4. The carpal tunnel is composed of the osseous floor of the carpal bones and the fibrous roof of the transverse carpal ligament (anterior superior view). Several tendons, including the median nerve run via this tube

Figure 4.5. Extensor retinaculum (posterior view)

Figure 4.6. The lumbrical muscles (palmar view)

Figure 5.1. Gait cycle terminology. Left or right steps comprise a gait cycle
Figure 5.2. The gait cycle's phases
Figure 5.3. The five elements of the stance phase
Figure 5.4. Swing phase
Figure 5.5. (A) Midswing; (B) midswing period (RLA)
Figure 5.6. Throughout the gait cycle, the body's center of gravity moves vertically
Figure 5.7. Lateral pelvic tilt
Figure 5.8. Gluteus medius gait
Figure 5.9. Gait resulting from quadriceps weakness
Figure 5.10. (A) The lumbar lordosis is pushed down and the pelvis is tilted posteriorly to achieve the leg swing forward position. (B) By raising lumbar lordosis and tilting the pelvis anteriorly, the leg swings backwards
Figure 6.1. The anterior-posterior curves of the vertebral column are seen
Figure 6.2. Neck and trunk movements
Figure 6.3. The body landmarks of the front and posterior halves of a typical vertebra
Figure 6.4. The costal facets (rib attachments) of the thoracic vertebrae are seen
Figure 6.5. Depicts the connection between C1 and C2, illustrating the three atlantoaxial connectors
Figure 6.6. The kind of motion is determined by the alignment of the facet joints
Figure 6.7. The sternocleidomastoid muscle
Figure 6.8. Depicts the three scalene muscle components
Figure 7.1. The bones of the lower extremities
Figure 7.2. The hip joint (anterior view)
Figure 7.3. The bones of the pelvis (anterior view)
Figure 7.4. Right hip bone
Figure 7.5. The hip joint capsule (anterior view)
Figure 7.6. The ischiofemoral, pubofemoral, and iliofemoral ligaments provide support for the hip joint capsule
Figure 7.7. The inguinal ligament (anterior view)
Figure 7.8. The three adductor muscles (anterior view)
Figure 8.1. Interosseous membrane and leg bones
Figure 8.2. Lateral view of right leg. Notice the fibula's posterior position
Figure 8.3. Skeletal components of the left foot (medial, lateral, and superior views)
Figure 8.4. The foot's functional areas (superior view)
Figure 8.5. Ankle joint and motions of the foot

Figure 8.6. Posterior view of ankle joint

Figure 8.7. Phalanges of the foot's joints

Figure 8.8. The kinematic axes of the ankle joint. (a) Superior view; (b) anterior view

Figure 8.9. Ligaments of the right ankle's lateral aspect

Figure 8.10. Muscles of the fourth (deepest) layer of the right foot's plantar surface. (a) Plantar interossei; (b) dorsal interossei

LIST OF ABBREVIATIONS

AP	anterior-posterior
ASIS	anterior superior iliac spine
CMC	carpometacarpal
CNS	central nervous system
DIP	distal interphalangeal
EMG	electromyography
IP	interphalangeal
MCP	metacarpophalangeal
ML	medial-lateral
MTP	metatarsophalangeal
MVC	maximal voluntary contraction
PCSOs	physiological cross-sectional areas
PIIS	posterior inferior iliac spine
PIP	proximal interphalangeal
PSIS	posterior superior iliac spine
RLA	Rancho Los Amigos
ROM	ranges of motion
SI	superior-inferior

ABSTRACT

In this modern era, people have adequate access to use a machine to decrease their physical work and increase their time values. But the environment and habits around us are so difficult that they can increase the physical and mental burden of stress.

The majority of times, it has been seen that issues regarding joints and bones are so much problematic from poor to rich people that they cannot live their personal as well as professional life. So, to manage this type of issue, one has to excel in their knowledge regarding Kinesiology.

This book "Introduction of Kinesiology" can provide us a basic knowledge regarding muscles, bones, joints, and the formation of joints.

This literature also provides us with physiological as well as pathological problems of joints. Functions of joints and their clinical assessment in the form of clinical examinations are also explained here in this work. At last normal gait and issues regarding gait are also explained in brief.

So, from the learning of literature explained above, one can provide valuable treatment and management to the desired ones. This literature can be used not only by a physician or a health worker but also used by a layman who wants to go for helping hands to others.

PREFACE

Most living species rely on movement to survive. We recognize that mobility might be hampered for a variety of reasons, including injury or discomfort. We also see a shared goal to better or optimize a movement. Normal movement is the consequence of a complex interplay between the neurological, muscular, and skeletal systems, as well as other systems such as the respiratory, cardiovascular, integumentary, and immunological systems. Movement can be called a "system" when seen as a whole. To create movement, the Human Movement System combines the activities of the other physiologic systems. Kinesiology is the basic science for understanding the movement system. It is the scientific study of the movement of the human body or its components. A solid basis in Kinesiology is required for the restoration of normal movement, the decrease of movement dysfunction, and the optimization of movement.

From the time of Aristotle (384–322 BC), Kinesiology focused on anatomy and how muscles and joints were engaged in various movements. One can still study Kinesiology in the university's Physical Education department and apply it to athletes, post-injury rehab, etc. First used by R W Lovett, a Boston Orthopedic surgeon, to assess polio and nerve damage patients' limitations (1932). Henry and Florence Kendall's Muscles: Testing and Function (1949) popularized Kinesiology muscle testing. In 1964, Dr. George Goodheart, a chiropractor in Detroit, Michigan, noticed that many chiropractic adjustments could not ease spasming muscles. He noted that when a muscle spasms (producing pain), it is because the opposite muscle is weak. So, if one upper shoulder (upper trapezius muscle) hurt, it was because the other was too weak.

Goodheart looked for strategies to balance his muscles. Massaging reflex sites found by an osteopath in 1931 helped. Goodheart was able to correlate muscles to organs because Chapman had linked reflex sites to organs. Lightly holding reflex spots, identified by a chiropractor in the 1930s, strengthens muscles. Bennett linked reflex sites to organs; Goodheart did the same for muscles. Tracing acupuncture meridians is another approach to strengthen weak muscles, and acupuncturists have linked them with organ functioning. Goodheart sometimes strengthened a muscle by stroking its ends. Goodheart's brilliance was that he would try everything until he found a solution if a muscle would not balance. Gradually, he created a system for quickly realigning the body by balancing muscles. The medical community has neglected to acknowledge muscle-organ connections. However, Dr. Goodheart showed that balancing muscles can increase organ performance. This approach became Applied Kinesiology, which can only be utilized by licensed physicians – mostly chiropractors, but also medical doctors, psychiatrists, etc.

A California chiropractor, Dr. John Thie, noticed that much of this content could be taught to laypeople and health professionals without a license to diagnose. Dr. Thie and colleagues published the Touch for Health guidebook in 1973 and began teaching laypeople how to balance the body's muscles and meridian energy lines by massaging or holding certain reflexes, tracing meridians, and working with nutrients, sound, color, emotions, etc.

This book (Introduction to Kinesiology) is divided into eight chapters. The first chapter introduces the readers with the fundamentals of Kinesiology. Chapter 2 explain the Kinesiology of the shoulder complex. Chapter 3 thoroughly discusses the Kinesiology of Elbow units.

Chapter 4 introduces the readers with the Kinesiology of the Hand. Chapter 5 focuses on the Kinesiology of Gait. Chapter 6 illustrates the phenomena of Kinesiology in the neck and trunk. Chapter 7 describes the Kinesiology in Hip Joints' modern applications of proteomics in various fields. Finally, Chapter 8 focuses on the Kinesiology of Ankle Joint and Foot recent.

The book gives a good understanding of the many aspects of Kinesiology. The work succeeds in delivering material in a way that familiarizes the unfamiliar reader with essential Kinesiology ideas and methods. Students in medical and other interdisciplinary disciplines will benefit from an introduction to Kinesiology.

CHAPTER 1

FUNDAMENTALS OF KINESIOLOGY

CONTENTS

1.1. Introduction 2
1.2. History of Kinesiology 4
1.3. Descriptive Terminology 9
1.4. Segments of the Body 12
1.5. Types of Motion 13
1.6. Joint Movements (Osteokinematics) 16
1.7. Kinesiology of Testing Muscles 22
1.8. Development of Traditional Kinesiology 24
References 26

1.1. INTRODUCTION

Kinesiology is the study of movement. However, this concept is too broad to be particularly useful. Kinesiology is a branch of science that connects physiology, anatomy, geometry, and physics to human movement. As a result, Kinesiology employs mechanics, musculoskeletal anatomy, and neuromuscular physiology concepts (Enoka, 1988; Jonsson, 1978; Simeón and Monge, 2005).

Biomechanics is the study of mechanical concepts that have been directly linked to the body of a human. We should evaluate our biomechanical relationship with a racket, ball, prosthesis, crutch, or other implements since we can utilize one. This could entail examining the *static* (nonmoving) and *dynamic* (moving) *systems* involved in various activities. Kinematics and Kinetics and kinematics are two types of dynamic systems (Frost, 2013; McAfee, 2014; Neumann, 2010a; Perry, 1974; Tomchuk, 2010; Twietmeyer, 2012). Kinetics is the study of the forces that cause movement, wherein kinematics is the study of the space, mass, and time of a moving system.

The musculoskeletal anatomy elements, which have been regarded as the key to understanding and implementing the other elements, would receive the most attention in this text. *Kinesiology* causes several students to have unfavorable sentiments. Their minds go blank and their eyes glaze over. Perhaps they believe that mass memorizing is their only hope, dependent upon previous experience using anatomy. However, this might be a difficult task with no long-term memory benefit (Neuman, 2010, b, 2016; Thompson, 2003).

Retain a certain fundamental principle in mind as you go through this chapter. Firstly, the human body is extremely well-organized. There have been exceptions to everything in life. The reasoning behind such exceptions is often obvious, and sometimes it is only clear to a higher being. In either scenario, make a note of the exception and carry on. Secondly, rigid memorizing is unnecessary if you have a strong grasp of specific terms and may comprehend the notion or characteristic. If you understand which the patella is and what structures surround it, for instance, you may properly explain its location utilizing your own words. To be accurate, you do not need to remember someone else's words (Donald and Shimada; Neumann 3rd; Neumann, 2010).

If you recognize (i) what movements a specific joint permit; (ii) that such a muscle should span a specific joint surface to create a motion; and (iii) what that muscle's line of pull is, you will be able to determine the exact action(s) of that muscle. For instance, (a) the elbow can only extend and flex; (b) to extend and flex the elbow, a muscle should span it posteriorly and anteriorly; (c) the biceps brachii is a vertical muscle located on the arm's anterior region; (d) as a result, the bicep flexes the elbow. (Explanation regarding movements and muscles used: the elbow can only extend and flex when a muscle span it posteriorly and anteriorly. The biceps brachii is a vertical muscle located on the arm's anterior region; as a result, the bicep flexes the elbow). Yes, even simple mortals may comprehend Kinesiology. Its research may even be pleasurable. Although, a note of caution is necessary: It is preferable to study in little increments multiple times a week, rather than studying for a long duration in one sitting before the exam (Beazell et al., 2012; Bell, Guskiewicz, Clark, and Padua, 2011; Biz et al., 2022; Gribble et al., 2013; Jackson, Simon, and Docherty, 2016; Lee and Lee, 2017; Paulovich, 2018).

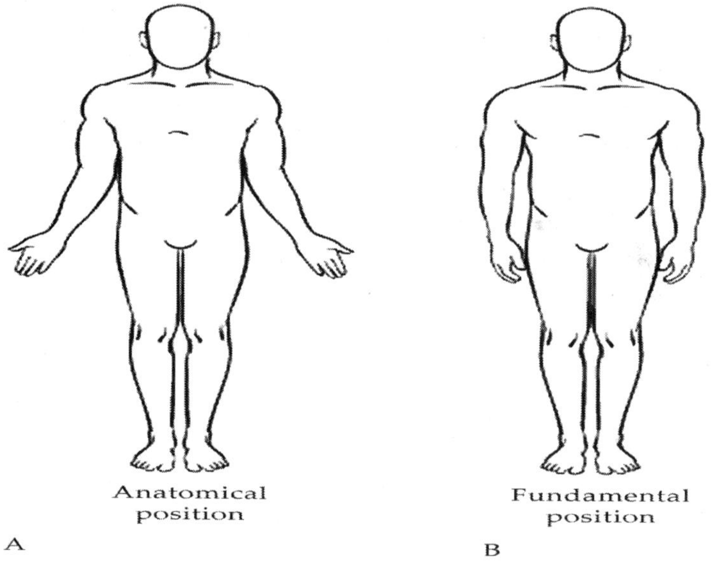

Figure 1.1. Descriptive positions.

Source: https://www.tamiapland.com/blog/2018/8/19/kinesiology-lingo-movements-of-the-body.

1.2. HISTORY OF KINESIOLOGY

Kinesiology (from Greek kinesis, motion) (Greek "kinesis" meaning 'motion') originated as a study of animal and human motion in ancient times. This initial, traditional kind of Kinesiology (biomechanics) has developed a vast amount of information about how nerves drive muscles to operate on bones to generate movement and posture throughout many years. Kinesiology, such as physiotherapy, is a therapeutic profession having a lengthy history. Medical testing of muscle was used in biomechanics even before Kinesiology became popular (Yin and Wang, 2020).

Kinesiology's biomechanics concepts (like using the least amount of force to achieve the best result) are effectively used in several ergonomic challenges in medicine, sports, and industry. The usage of biomechanics in the industry has led to the creation of "user-friendly" chairs, instruments, workstations, and other items. It has sparked the creation of ergonomic work procedures (for example, how to move big objects safely) that lead to fewer injuries and more efficiency. Athletes work alongside kinesiologists to learn and perform the motions needed by their sport more effectively and productively. Biomechanics concepts have numerous applications in medicine, such as the creation of more efficient enhanced techniques and the design of prosthetic joints.

Biomechanics, often known as "Conventional" Kinesiology, has a long history that dates back millions of years and continues to this day. Kinesiology, on the other hand, started in 1964 with the studies of George J. Goodheart, Jr., an American chiropractor. Many of the diagnostic tools employed nowadays in this comparatively recent subject are derived from his outstanding talents of curiosity, observation, and passion to explore the reasons for what he observed and the discoveries he made. Dr. George J. Goodheart developed Kinesiology after analyzing his daily chiropractic practice (Atwater, 1980; Hamill, Knutzen, and Derrick, 2021).

In his practice, he found that the use of conventional chiropractic techniques was, for the most part, successful. On the other hand, he was deeply troubled by the fact that he lacked the skills necessary to accurately diagnose the odd collection of contradictories (or merely perplexing) symptoms. And because he lacked a proper diagnosis, he had been helpless when it came to developing appropriate remedies (Kaptchuk and Eisenberg, 1998; Leach, 2004). When perplexed by a patient's peculiar diagnostic findings and symptoms, Dr Goodheart repeatedly posed the maturity-level question of the scientist or researcher: "Why is this?"

Goodheart extensively explored the physiology and anatomy associated with his patient's difficulties in his quest for explanations that would lead to efficient remedial methods. This understanding frequently leads him to potential interventions. His treatment and theory of reactive muscles, as well as his appropriate treatment for sustained muscular usage, are instances of novel procedures he devised. Each of these strategies has been discussed in greater depth in the following section. Goodheart went beyond the confines of his professional chiropractic education to investigate the ideas of other inventive scientists and healers. He examined the traditional wisdom and research results of numerous different therapeutic systems (lymphatic drainage, Chinese acupuncture, neurology, nutrition, etc.), (lymphatic drainage, Chinese acupuncture, neurology, nutrition, etc.) – and then incorporated them into Kinesiology. He explored numerous other forms of treatment, including Bennett's reflexes, Chapman's reflexes, synchronization of pulses at reflex spots, and so on. When he was unable to address a perplexing diagnostic issue with the methods he already possessed, he explored even the most unorthodox methods (Gin and Green, 1997; Goodheart and Cox, 1994). The majority of approaches utilized in Kinesiology nowadays are the result of his unconventionally open-minded study of methodologies. Goodheart discovered that marked the start of Kinesiology in 1964.

As a chiropractor, he believed that treating structural imbalances in the body (posture issues, incorrect bone arrangement, etc.), would minimize or remove the majority of health issues. The objective of chiropractic care is to restore structural balance and proper postural orientation of the body's components. However, structural equilibrium cannot be achieved when muscles have been excessively tense or lax (Figure 1.2).

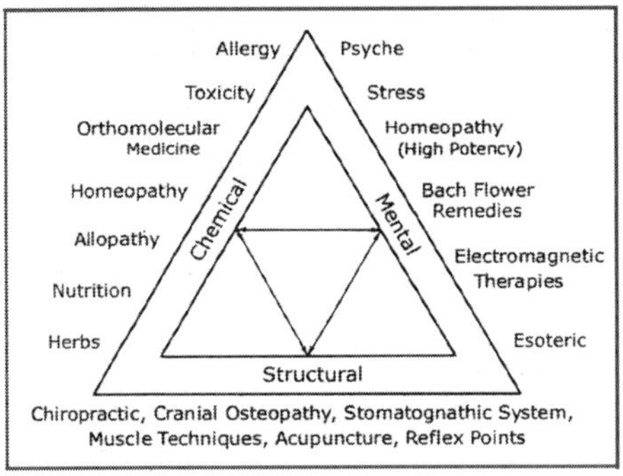

Figure 1.2. The triad of health.

Source: https://www.synergychiros.com/blog/post/the-triad-of-health.html.

Goodheart treated inadequately over many months a patient whose initial symptom was one shoulder blade that protruded away from his back. He recalled reading the classic Muscles: Testing and Function by Kendall and Kendall about a muscle that dragged the shoulder blade down onto the back. He evaluated the serratus anticus muscle using the muscular testing approach provided by Kendall and Kendall (known as the anterior serratus). Just the side in which the shoulder blade protruded tested weak.

Because of its "toothed" structure, the serratus muscle is nicknamed "serrated" like a saw. It joins the top eight or nine ribs to the scapula, or shoulder blade, on the inner vertebral side. The muscle on the weak-testing side was not less developed, and Goodheart could find no alternative explanation for only one test weakness. With his fingertips, he explored the weak-testing muscle and discovered painful tiny lumps (nodules) in which the serratus anticus tendons link to the ribs. Whenever he massaged one of such nodules hard enough, it vanished. He massaged all of such nodules hard as an experiment and discovered an instant rise in the muscle's "test strength" when retested. Goodheart was inspired by this finding and utilized Kendall and Kendall's book to educate himself on how and when to muscle-test a variety of additional muscles. This had been Kinesiology's 1st finding and the start of a long and profitable study project. The "origin-insertion technique" was named after the surprise finding that a weak-testing muscle can be persuaded to test strong by massaging its upper end, in which its tendons join to bone. Several chiropractors started to employ manual muscle testing to measure structural balance, which is the purpose of chiropractic because this approach proved to be effective in restoring muscular balance (and consequently structural balance). Several additional health issues vanished without therapy when the origin-insertion massage strengthened weak-testing muscles. This supported the core chiropractic notion that structural balance has an impact on all facets of health (Goodheart, Kazdin, and Sternberg, 2006).

The origin-insertion approach, on the other hand, is frequently unable to improve weak-testing muscles and restore muscular balance. Muscle-building workouts were also ineffective. These workouts, which were developed particularly to improve the weak-testing muscle, increased the muscle's bulk and weight-bearing capability; however, the muscle is still "muscle-tested" weak. Other variables beyond pure physical strength had been at play, and they required to be discovered (Hamilton et al., 2018; Lutgendorf et al., 2011). Additional research by Goodheart demonstrated that muscular disparities could be caused by issues not only in the muscle's origin-insertion area, but also in any of the three areas depicted by the chiropractic "triad of health," that is, dysfunction can be caused by chemical, structural, and/or mental issues.

In the Kinesiology examination, the interplay of the three sides of the triad of health is an essential and extremely valuable idea. There are several popular instances of how one side of the triangle can influence the other. Several meals or drugs, for instance, can create mental problems. In preparations for fight or flight, fear (psychological) leads to the secretion of adrenaline, and this raises the tension in skeletal muscles (structural). Neck tension (structural) can lead to serious headaches and sadness (mental). Emotional issues (mental) can lead to a build-up of acid in the stomach (chemical), causing pain and causing the patient to lean down and forward (structural).

Many healing professionals are expertise in one aspect of the health triad. Massage, chiropractic, surgery, dentistry, and osteopathy are examples of structural therapies. Nutrition and medicine are examples of chemical therapies. Psychology and counseling are among the Mental treatments. Specialists in such systems are rarely well-trained in coping with issues that affect the other two sides of the health triad. When dealing with patients, professionals automatically employ notions and approaches that they have been familiar with (Nigg, MacIntosh, and Mester, 2000).

However, counseling is unlikely to alleviate a headache if its primary cause is dietary. To visually demonstrate this argument, if one's sole tool is a hammer, the entire universe appears to be made of nails. All such professionals require improved methods for detecting the reasons for their patients' issues and deciding which therapies are efficient. Typically, when a health issue persists over a lengthy period (becomes chronic), all three aspects of the health triangle are implicated. Moreover, the issue that takes a patient to a therapist is frequently not the major issue, but instead a reaction to one of the other sides of the triangle. As long as the core condition remains undiagnosed and untreated, the same symptoms can reappear, and more secondary issues can occur.

Goodheart realized that he was required to broaden his scope of the investigation to best serve his patients. He understood the importance of being able to assess and treat issues on all three sides of the triad of health through his chiropractic studies. To accomplish so, he looked into the muscle testing technique's ability to test all three components exhaustively.

He was satisfied after thorough research that muscle testing was effective in assessing all of the components that impact health.

He discovered that:

- Certain health issues can cause certain muscles to test feeble;
- A muscle that tests feeble as a result of a health concern might be utilized as a marker to assess therapy options;
- Treatments that strengthen the muscle test can have a favorable impact on the health issue.

Goodheart's elasticity of mind is remarkable, especially when considering his chiropractic training. Even so, he had already been concentrated upon the restricting conceptions of his particular field of study, as any professional would be. In his pursuit of the chiropractic aim of structural balance, Goodheart studied a broad range of treatment techniques with unbridled enthusiasm. Any method that led to the improvement of a weak testing muscle was extensively investigated by him.

Patients having similar symptoms may need quite distinct treatments. Several effective therapies evaluated by Goodheart were established but were seldom employed owing to a shortage of diagnostic procedures that might determine whether a particular intervention will be beneficial. Muscle testing gave him the assessment tool he required to pick from the numerous viable therapies for every issue. Muscle testing gives a direct technique for investigating the impact of any sort of treatment modality on the body as it employs the patient's own body as the tool for diagnosis. Muscle testing, according to Goodheart, is the most straightforward technique for determining which treatment is most suitable for every patient's needs.

Acquired, altered, improved, and standardized several strategies beneficial in the improvement of weak-testing muscles for his usage and the advantage of other therapists. His work is responsible for the majority of the approaches that are recognized and employed nowadays. The research of Goodheart is notable for its intuitive conceptual jumps. He discovered, for instance, that a proper treatment measure nearly always quickly restores a weak-testing muscle to a strong-testing muscle (Heggers et al., 2005).

He then instinctively concluded that muscle testing could be utilized to assess the efficiency of any therapy after it had been administered. Further investigation revealed that the inspiration was correct. He created the rule that following a therapy, muscle testing may be utilized to assess whether the treatment was beneficial.

Today's Kinesiology practitioners stand on the shoulders of giants like Goodheart who came before them. From this vantage point, we may falsely believe that these intuitive leaps are clear. The task is to be such a giant and to uncover some of the comparable "jumps" of revelation that remain undiscovered in this nascent field of inquiry.

Standard medical muscle tests of biomechanics are used by a wide range of different types of health professionals who have received training in AK to directly evaluate the normal function of the muscles and the nervous system. Later on in this book, a more in-depth discussion on muscle testing is included. A concise explanation of muscle testing as it is carried out in Kinesiology is provided below as a means of providing a first introduction.

1.3. DESCRIPTIVE TERMINOLOGY

The human body is dynamic and in continual motion. It is prone to frequent positional changes. The relationships between the different physical components also alter. To explain the orientation of the human body, this is required to choose an arbitrary position through which to explain the movement or placement of structures. This is referred to as the anatomical posture (Figure 1.1(A)) and is characterized by the human body standing erect with the eyes looking ahead, the feet parallel to one another and relatively close together, and the arms resting at the sides of the body with the palms facing forward. Whereas the posture of the hands and forearm is not natural, it permits complete explanation (Carini et al., 2017; Chi and Kennon, 2006).

The basic posture (Figure 1.1(B)) is similar to the anatomical position, with the exception that the palms are towards the sides of the body. When addressing rotation of the upper extremities, this posture is frequently employed. The positions of a structure and its relationship to other structures have been described using specific terminology (Figure 1.3). A position closer to the midline is referred to as medial, whereas a position away from the midline is referred to as lateral. In the ulna, for instance, the radius is on the lateral side, whereas it is on the medial side of the forearm (Cuccia and Caradonna, 2009).

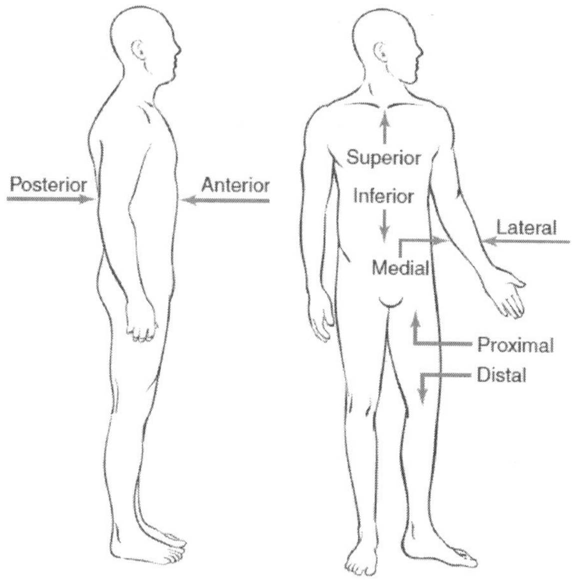

Figure 1.3. Descriptive terminology.

Source: https://www.slideshare.net/persaud_dan/foundational-kinesiology-module-1-basic-concepts-jan-2011.

The front of the body or a location nearer to the front is referred to as anterior. The back of the body or a location that is towards the back is referred to as posterior. The sternum, for instance, is located anterior to the chest wall. Positions on the extremities have been described as proximal and distal. Distal indicates far from the trunk, while Proximal implies toward the trunk. The humeral head, for instance, is found near the proximal end of the humerus. The elbow is near the wrist but far from the shoulder.

Superior refers to the top surface of an organ or structure, or the placement of a bodily component being above another. Inferior describes a bodily portion that is under another or the lowest surface of a structure or organ. The sternum's body, for instance, is higher than the xiphoid process but lower than the manubrium. A location or structure near the head is sometimes referred to as cephalad or cranial (from the word root cephal, indicating "head"). A location or structure that is closer to the feet is said to be caudal (Sherman et al., 2015). The word caudal comes from the word root cauda, which means "tail." A good illustration of this is the cauda equina, which translates to "horse's tail" and refers to the bundle of spinal nerve roots that descend from the bottom of the spinal cord.

Similarly, to the phrases ventral and dorsal, caudal, and cranial have been ideally utilized to define places on a quadruped (an animal having four legs). Humans are bipeds or creatures with two legs (Lafon, Smith, and Beillas, 2010). Figure 1.4 demonstrates that if the dog were to rise on its back legs, cranial will become superior, dorsal will become posterior, etc. Based on its comparative depth, a structure can be characterized as deep or superficial. In discussing the strata of the abdominal muscles, for instance, the outer oblique is deep relative to the rectus abdominis but superficial relative to the inner oblique. Another illustration is the description of the scalp as being superficial to the skull.

The phrases prone and supine have been utilized to express the body's posture when resting flat. When in the supine position, a person's face, or anterior surface, is facing upward. The face, or anterior surface, of a person in the prone posture is angled downward (Figure 1.6 depicts a child prone on the sled). Bilateral consists of two sides, or even both. For instance, bilateral above-knee amputations are related to the amputation of both the left or right legs above the knee. The opposing side is referred to as the contralateral. A person who suffers a stroke impacting the right side of the brain, for instance, may experience contralateral paralysis of the left arm and leg. Alternatively, an ipsilateral describes a similar side of the body (Zampini, Harris, and Spence, 2005).

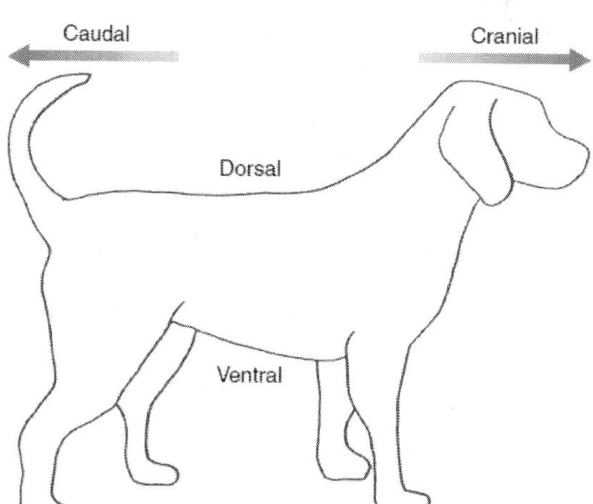

Figure 1.4. A quadruped's descriptive terminology.

Source: https://www.physio-pedia.com/Introduction_to_Quadruped_Anatomical_Terminology.

1.4. SEGMENTS OF THE BODY

As per the bones, the body is split into sections (Figure 1.5). The arm is the bone (humerus) that connects the shoulder to the elbow joint in the upper extremity. Between the wrist and the elbow lies the forearm (ulna and radius). The wrist is distal to the hand. Three comparable fragments make up the lower extremity. Between the knee joint and the hip is the thigh (femur). Between the ankle joint and the knee lies the leg (fibula and tibia), and the foot is distal to the ankle (Clauser, McConville, and Young, 1969; Dempster and Gaughran, 1967).

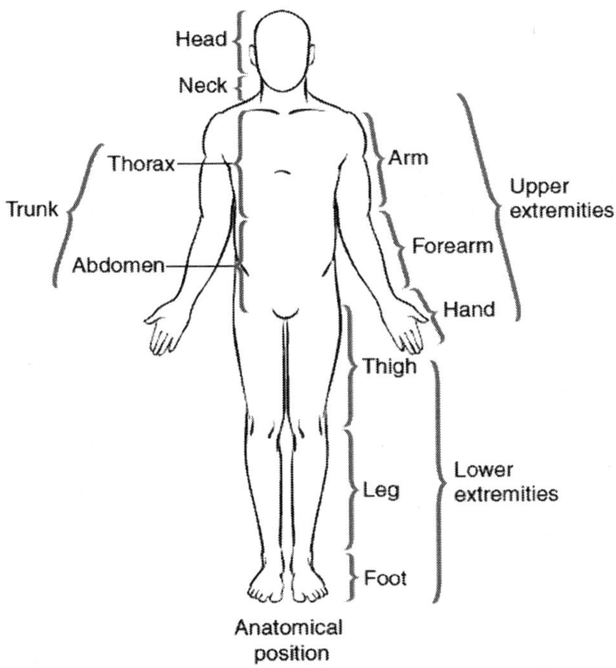

Figure 1.5. Body segments.

Source: https://www.researchgate.net/figure/Body-segments-and-examples-of-thermal-nodes_fig1_331524583.

Two components comprise the trunk: the abdomen and the thorax. The chest or thorax consists primarily of ribs and thoracic vertebrae. The lower trunk or abdomen consists of the stomach, the pelvis, and the majority of the lumbar vertebrae. The neck (cervical vertebrae) and head (skull) are different portions. Arthrokinematic motion refers to the surface motion of a joint in connection to the motion of a body part. For instance, the surface of the proximal end of the humerus goes downward, whereas the arm segment travels upward. Body segments are occasionally utilized to depict joint motion. For instance, flexion of the shoulder happens, not the flexion of the arm. The movement happens at the joint (shoulder), and the body component (arm) merely accompanies it! The exception to this generalization is the forearm. It is a part of the body that also works as a joint. Technically, joint mobility happens at the distal and proximal radioulnar joints, although this is commonly referred to as supination and pronation of the forearm.

1.5. TYPES OF MOTION

Linear motion, also known as translatory motion, happens in a relatively straight line between two points. All of the object's components move at the same distance, in a similar direction, and at the same rate. Rectilinear motion is a movement that happens in a straight line, including a youngster sliding down a hill (Figure 1.6), a sailboarder crossing the water, or a baseball player going from home plate to 1st base (Aminian and Najafi, 2004; McDonnell et al., 2009).

Figure 1.6. Curvilinear motion.

Source: https://www.real-world-physics-problems.com/curvilinear-motion.html.

Curvilinear motion happens when change happens along a curved pathway which is not necessarily circular. Curvilinear motion is the pathway a diver makes after departing the diving board and before going into the water. Figure 1.7 depicts a skier's curvilinear pathway as they down a ski slope. The route of a thrown ball, a javelin hurled along a field, or the Earth's orbit around the sun is all instances of curvilinear motion. Angular motion often referred to as rotary motion, is the motion of a body around a fixed point (Figure 1.8).

All of the object's parts travel in a similar direction, at the same angle, and at the same time, but they do not move the same distance. The foot goes farther across space as compared to the leg or ankle when a person flexes his knee.

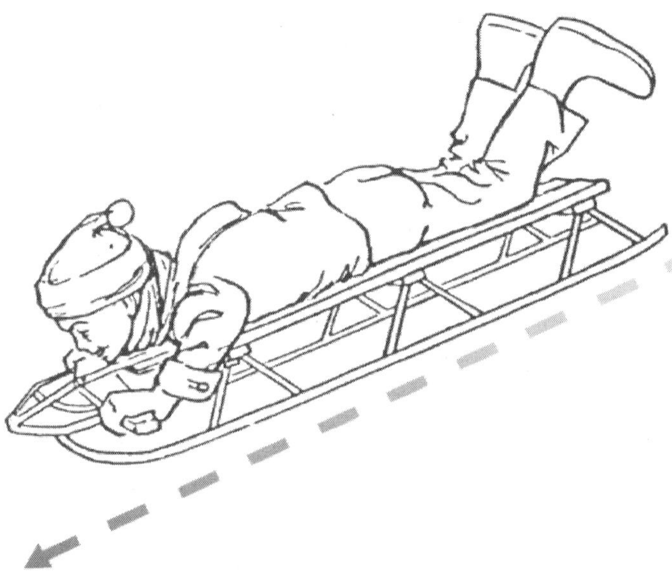

Figure 1.7. Rectilinear motion.

Source: https://studiousguy.com/linear-motion-examples/.

This is not unusual for both kinds of motion to take place simultaneously, with the total item traveling linearly and the individual pieces moving in an angular manner (Loper, Mahmood, and Black, 2014; McDonnell et al., 2009; Slaughter, Heron, and Sim, 2002). In Figure 1.9, the skateboarder's entire body glides along the street (linear motion), whereas the ankle, knee, and hip of the "pushing" leg-spin about its axes (angular motion). Walking is another form of linked motion. The entire body shows linear motion when walking between point A and point B, whereas the ankles, knees, and hips demonstrate angular top extremity joints. The route of the ball is curvilinear.

The movement inside the body is typically angular, but movement external side of the body is typically linear.

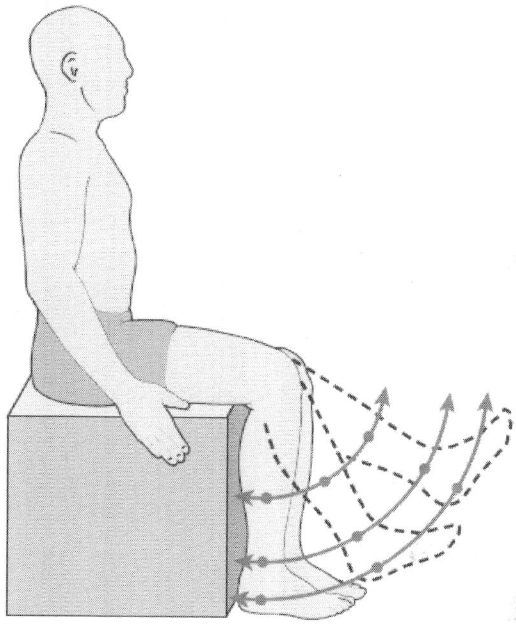

Figure 1.8. Angular motion.

Source: https://www.teachpe.com/biomechanics/angular-motion.

There are exceptions to this assumption. For instance, the scapula's depression or elevation and retraction or protraction movements are largely linear. The clavicle, which is linked to the scapula, moves in an angular fashion, and receives its angular motion from the sternoclavicular joint.

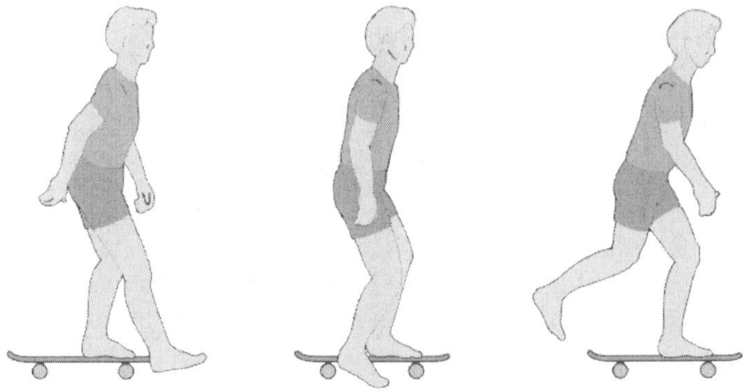

Figure 1.9. Motion that combines angular and linear motion.

Source: https://www.nbcorporation.com/can-bearing-used-linear-rotational-motion/.

1.6. JOINT MOVEMENTS (OSTEOKINEMATICS)

Joints can move in numerous directions. Movement happens along joint axes as well as via joint planes, as would be explained. The terminologies listed below have been utilized to explain the different joint motions that happen at synovial joints (Figure 1.10). Synovial joints are mobile joints in which the majority of joint motion happens. This form of joint motion is also known as Osteokinematics, which studies the link between the mobility of bones along with a joint axis (for example, humerus moving on scapula) and joint surface movement (humeral head's motion inside glenoid fossa of scapula) (Lee et al., 2017).

Flexion is the motion of one bone bending over another, putting the two fragments closer together and increasing the joint angle. Typically, this happens between the anterior surfaces of articulating bones, and the surfaces approach one another. In the situation of the neck, flexion is a "bowing down" movement wherein the head travels toward the anterior chest (Figure 1.10(A)). By flexing the elbow, the arm and forearm move near one another. Consequently, at the knee, the posterior surfaces (leg and thigh) move toward one another, causing flexion. When the lower extremity is the moving component, hip flexion causes the thigh to move anteriorly toward the trunk. This occurs whenever the lower extremity is in motion (Smith, Vitharana, Wallis, and Vicenzino, 2020). When the lower extremities are held in place, and the trunk is allowed to become the moveable component, the trunk will flex.

Depending on your point of view, flexion may lead to an increase or reduction in joint angle. When measuring elbow flexion with goniometry, the anatomical position (maximum extension) is regarded as zero. The angle of flexion approaches 180°. In this situation, flexion will indicate a rise in joint angle (Figure 1.10(D)). In other contexts, flexion starts at 180° (maximum extension) and progresses toward 0°; therefore, this is a reduction in the joint angle. In contrast, extension is the motion of a single bone straightening out from another, which increases the joint angle. This action often returns the flexed bodily component to its normal anatomical position (Figure 1.10(B), (E)). The joint surfaces tend to separate (Ko). When the head goes up and away from the chest and the thigh drifts away from the trunk and returns to its natural place, extension happens. The continuance of extension further than the anatomical position is known as hyperextension (Figure 1.10(C)). Trunk, shoulder, neck, and hip hyperextension is possible. Flexion at the wrist is referred to as palmar flexion (Figure 1.10(F)), whereas flexion at the ankle is referred to as plantar flexion (Figure 1.10(H)). Dorsiflexion could refer to the extension of the wrist and ankle joints (Figure 1.10(G), (I)).

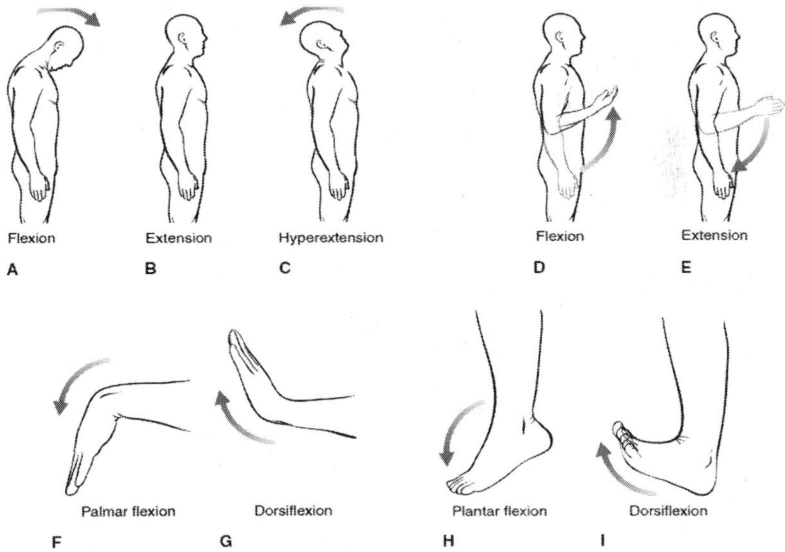

Figure 1.10. Flexion and extension of the joints.

Source: https://teachmeanatomy.info/the-basics/anatomical-terminology/terms-of-movement/.

Abduction (Figure 1.11(A)) is the movement away from the body's midline, whereas adduction (Figure 1.11(B)) is the movement toward it. The shoulder and hip may adduct and abduct. The toes and fingers are the only exceptions to this midline definition. The middle finger serves as a reference point for the fingers. Abduction is the movement away from the middle finger (shown in Figure 1.11). It must be observed that the middle finger abducts (to the right) and the 2nd toe abducts (to the left) but does not adduct unless it is returning from abduction (Kage et al., 2020; Nirendan and Murugavel, 2019).

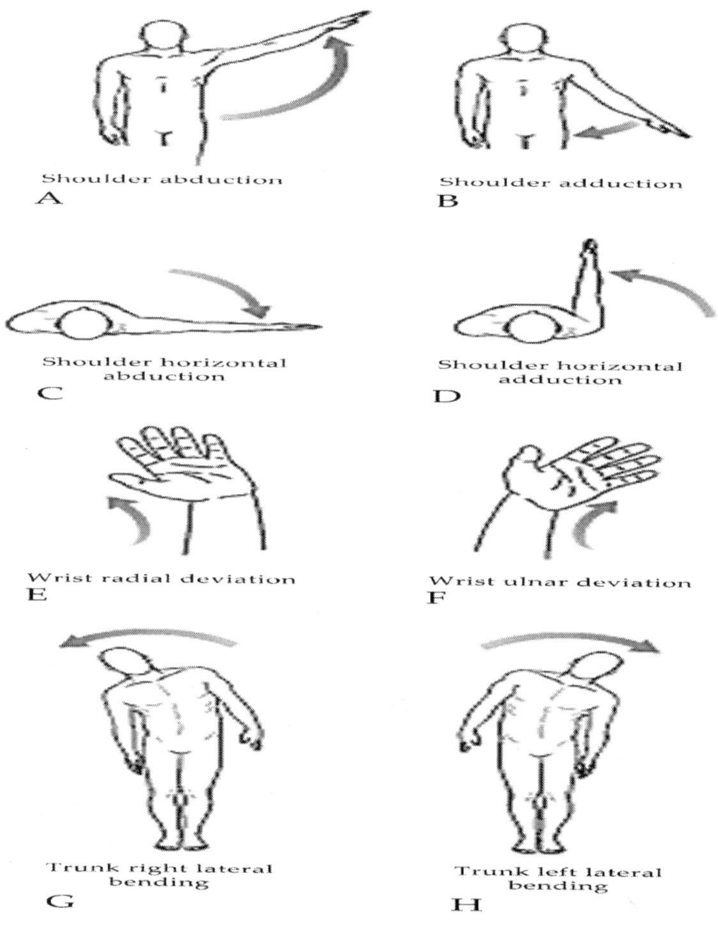

Figure 1.11. Joint motions of abduction and adduction.

Source: https://open.oregonstate.education/aandp/chapter/9-5-types-of-body-movements/.

Horizontal adduction and abduction are movements that cannot be performed due to anatomical constraints (Swanson and Frappier, 2001). They should be followed either by shoulder abduction or flexion to bring the arm to shoulder level. The backward movement of the shoulder (horizontal abduction) and the forward movement of the shoulder (horizontal adduction) are both possible from this position (Figure 1.11(D)). Same movements are possible at the hip, but the ranges of motion (ROM) are usually limited. Wrist adduction and abduction are most generally referred to as ulnar deviation and radial deviation (Neumann, 2010b).

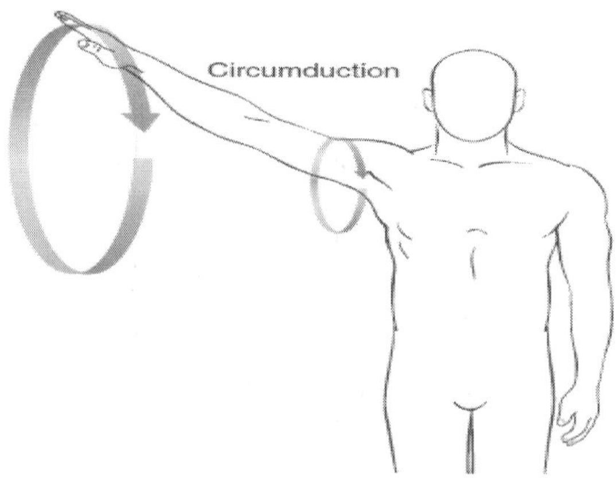

Figure 1.12. Circumduction motion.

Source: https://www.coursehero.com/study-guides/ap1/types-of-body-movements/.

Radial deviation occurs when the hand moves toward the thumb side or horizontally (Figure 1.11(E)). Ulnar deviation occurs when the hand shifts medially from its natural location toward the little finger side at the wrist (Figure 1.11(F)). The word lateral bending refers to the movement of the trunk sideways (Frost, 2013; Maugeri et al., 2020). The trunk may bend to the right or left horizontally (Figure 1.11(G) and (H)). Right lateral bending occurs when the right side of the trunk bends, shifting the shoulder toward the right hip. Similarly, the neck bends horizontally. This sideward movement is often referred to as lateral flexion. This phrase would not be utilized in this book since it is easily mistaken with flexion (Dishman, Washburn, and Schoeller, 2001; Lee, 1993; Vouloutsi et al., 2018).

Circumduction depicts a circular, cone-shaped arrangement of motion. There are four joint motions involved: (i) extension; (ii) flexion; (iii) adduction; and (iv) abduction. If the shoulder travels in a circle, for instance, the hand travels in a much large circle. The whole arm will rotate in a cone-shaped sequence of abduction, flexion, adduction, and extension, eventually returning to its original position (Figure 1.11). The movement of a bone or component around its longitudinal axis is known as rotation (Shorter, Wu, and Kuo, 2017). The medial rotation occurs when the anterior surface slides inward toward the midline (Figure 1.13(A)). Internal rotation is a term used to describe this (Salvia et al., 2000). In contrast, lateral rotation (Figure 1.13(B)) or external rotation occurs when the anterior surface rolls outward, away from the midline. The trunk and neck can be rotated to the left or right (Figure 1.13(C) and (D)). As you glance over your right shoulder, imagine the neck turning. This is referred to as "right neck rotation."

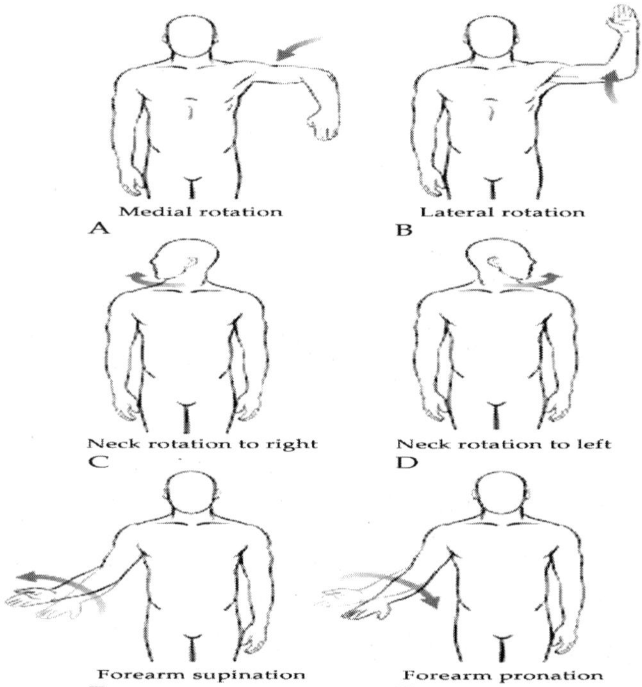

Figure 1.13. Joint rotation motions.

Source: https://www.pharmacy180.com/article/types-of-joint-movements-3510/.

Pronation and supination are two terms for forearm movement. The forearm is in supination in a natural position (Figure 1.13(E)). The hand palm is facing forward, or anteriorly. The palm is pointing backwards, or posteriorly, in pronation (Figure 1.13(F)). Supination occurs whenever the elbow remains flexed, whereas pronation occurs whenever the elbow becomes flexed. The terminology listed below has been utilized to define joint-specific movements. Eversion is the motion of the foot sole inwards to the ankle (Figure 1.14(A)), while inversion is the motion outward (Figure 1.14(B)). Protraction is primarily a linear movement far from the midline in a plane parallel to the ground (Figure 1.15(A)), whereas retraction is primarily a linear motion toward the midline in a similar plane (Figure 1.15(B)). Protraction of the shoulder girdle, as well as protraction of the jaw, pulls the scapula far from the midline, while retraction in both situations restores the body component to its natural position (Johansson, Backlin, and Burstedt, 1999; Kapandji, 2001).

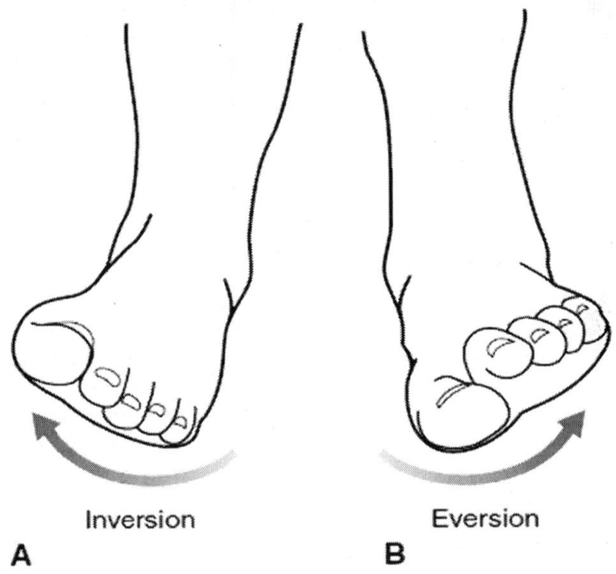

Figure 1.14. Inversion and eversion of left foot.

Source: https://anatomyzone.com/articles/inversion-foot/.

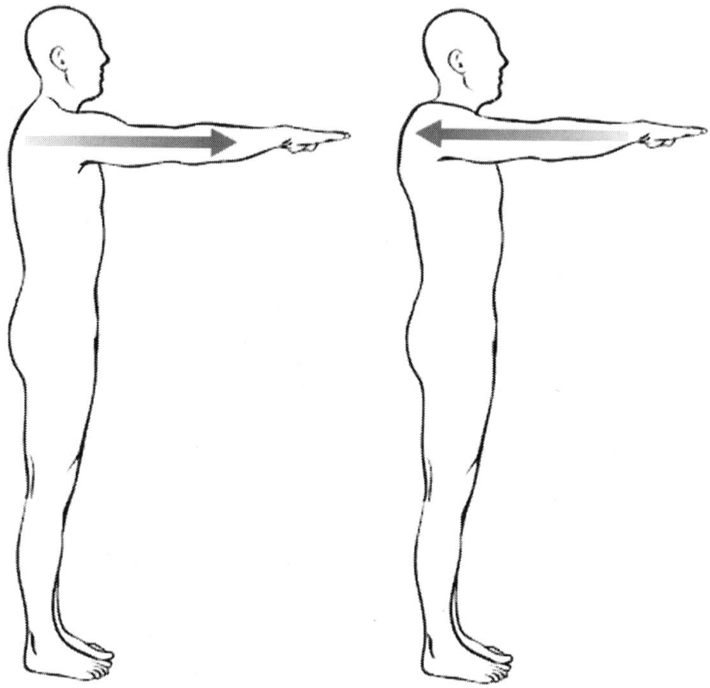

Figure 1.15. Protraction and retraction.

Source: https://commons.wikimedia.org/wiki/File:Protraction_Retraction.png.

1.7. KINESIOLOGY OF TESTING MUSCLES

The tendons at both ends of the majority of muscles have been linked to bones which intersect at a movable joint. Whenever muscles contract, they reduce in length. This shortening brings one of the connected bones closer to the other.

To prepare for the muscle test, the joint over which the muscle is connected is bent. This brings the muscle into a contracted state by shortening it. The examiner positions his hand to impede further muscle contractions.

The patient commences the test by gradually contracting the muscle against the examiner's immobile hand from zero to extreme force. Throughout this brief interval, the examiner applies an equivalent and opposing, gradually rising resistance to sustain the muscle test's initial posture (Bagheri et al., 2018). When the patient has maximally voluntarily contracted his muscle, the tester adds a bit of additional pressure. The entire testing procedure must not exceed two to three seconds. If the patient can hold the initial test posture while being subjected to a slight additional force, the muscle has "tested strong." If not, it "failed the test."

In the 1st portion of the muscle test, one evaluates the patient's resolve and capacity to contract the muscle forcefully. In the 2nd portion of the muscle test, one additionally evaluates the patient's nervous system's capacity to contract the muscle a bit more than the patient may do voluntarily. By using this approach, one is analyzing the functional integrity of the whole muscle circuit and the section of the nervous system that controls that muscle. This initial muscle test is conducted "in the clear," or without any additional stimuli. The muscle is contracted to the patient's maximum conscious capacity. After the patient has fully contracted the muscle and the examiner exerts extra pressure, the 2nd portion of the muscle test asks if the patient's neurological system may coordinate the muscle to contract slightly more than he is cognitively capable of.

In addition to "clear" muscle testing, as mentioned above, Kinesiology also employs "indicator" muscle testing. In this sort of muscle testing, a muscle that tested strongly in the clear is utilized as a signal for testing a different stimulus. Touching a part of the patient's body which is "disturbed" or malfunctioning due to injury, illness, etc., might offer additional stimulation. If performed when performing the test of an indicator muscle that recently tested strong, this stimulation can lead the muscle to test weak. Therapy localization refers to the stimulation delivered by the patient touching themselves. Several examiners touch the patient to therapy-localize, which is typically simpler, quicker, and yields similar outcomes. On occasion, although, the outcomes of therapeutic localization are distinct whenever the examiner touches the patient compared to when the patient touches a similar place of the body. For therapeutic localization, it is advised that the patients touch themselves. This is referred to as "challenge" when the patient is confronted with a stimulus other than touch, or when he or she does an activity, and the impact is then assessed through muscle testing.

The fact that almost all elements impacting health can be assessed utilizing an indicator muscle and treatment localization or challenge is part of Kinesiology's appeal. Standard muscle testing "in the clear" and indicator muscle testing of different stimuli are used by health professionals trained in Kinesiology procedures to assess the emotional or mental, structural, and biochemical processes of the human organism, as would be discussed later (Conable, 2010; Shih, Lee, and Chen, 2018).

Kinesiology is largely used for diagnosis. Whereas significant tools for the evaluation (diagnostic) of dysfunction had been created early in the field of AK, the majority of the therapies utilized in AK have come from other (often quite foreign) therapeutic fields. The practical benefit of AK, aside from its well-developed diagnostic tools, is that this may be determined which of several different therapeutic strategies would be the most beneficial for the unique difficulties of particular patients. As a result, before using any therapeutic procedure, the examiner may establish its relative strengths and so select the most appropriate therapy from a wide range of options (Hall, Lewith, Brien, and Little, 2008; Jacobs, Franks, and Gilman, 1984).

The diagnostic tools used by AK enable one to establish which bodily system is disrupted and which therapy modalities are suitable to correct the disruption. Interventions of all kinds (chemical, structural, nutritional, electromagnetic, mental, and so on) can be individually investigated before being used to cure a particular condition. The same procedures may be used after therapy to see if the treatment had been suitable, accurately applied, and successful (Lobbezoo, Van der Zaag, Visscher, and Naeije, 2004).

1.8. DEVELOPMENT OF TRADITIONAL KINESIOLOGY

Since the time of Aristotle (384–322 B.C.), commonly referred to as the "father of biology," the studies of motion (the traditional Kinesiology) have focused on mechanics and anatomy. Leonardo da Vinci's (1452–1519) studies on human anatomy and physiology are particularly popular (Ainsworth and Hooker, 2015; Lichtenstein et al., 2018).

This makes him one of the most famous pioneers in the field of Kinesiology, the study of motion. Mechanics is the field of physics concerned with the effects of forces and energy on things. The primary focus of earlier kinesiologists was the mechanical analysis of how muscles engage bones and joints to generate movement and posture. Ultimately, in the modern world, portraying the joints as fulcrums, the bones as levers, and the muscles as springs gave a straightforward mathematical description of body mechanics (Lowrie and Robinson, 2013).

These models give helpful insights into the motion of humans and other animals. Traditionally, Kinesiology was described as the study of the function and structure that cause human and animal movement. This field of study is now known as biomechanics and is often known as "conventional" Kinesiology.

After the extraordinary achievements of Leonardo da Vinci in the 14th century, classical Kinesiology made few advances for further than 200 years. Luigi Galvani discovered in 1780 that muscle contraction is caused by electric impulses, Kinesiology's progress resumed. Galvani administered a modest electrical voltage to the leg of a frog, which caused the leg's muscles to twitch. This led him to the right conclusion that electric impulses trigger muscle contraction (Braun, 1941; Sloane, 1952).

Before his time, this was believed that a muscle had its own volition. This mentality is still evident in statements like "the biceps act to move the wrist nearer the shoulder." Galvani's experiment revealed that electrical stimulation of muscles causes the contraction of the muscle and, by extension, bodily movement (Rosheim, 1997). With the finding that electric impulses are generated by nerves under the direction of the central nervous system (CNS) in live animals, the analysis of the function of nerves and the CNS (Neurophysiology) became an integral part of the study of movement (Kinesiology) (Elvan and Ozyurek, 2020).

REFERENCES

1. Ainsworth, B. E., & Hooker, S. P., (2015). The fusion of public health into Kinesiology. *Kinesiology Review, 4*(4), 322–328.
2. Aminian, K., & Najafi, B., (2004). Capturing human motion using body-fixed sensors: Outdoor measurement and clinical applications. *Computer Animation and Virtual Worlds, 15*(2), 79–94.
3. Atwater, A. E., (1980). Kinesiology/biomechanics: Perspectives and trends. *Research Quarterly for Exercise and Sport, 51*(1), 193–218.
4. Bagheri, R., Pourahmadi, M. R., Sarmadi, A. R., Takamjani, I. E., Torkaman, G., & Fazeli, S. H., (2018). What is the effect and mechanism of Kinesiology tape on muscle activity? *Journal of Bodywork and Movement Therapies, 22*(2), 266–275.
5. Beazell, J. R., Grindstaff, T. L., Sauer, L. D., Magrum, E. M., Ingersoll, C. D., & Hertel, J., (2012). Effects of a proximal or distal tibiofibular joint manipulation on ankle range of motion and functional outcomes in individuals with chronic ankle instability. *Journal of Orthopedic & Sports Physical Therapy, 42*(2), 125–134.
6. Bell, D. R., Guskiewicz, K. M., Clark, M. A., & Padua, D. A., (2011). Systematic review of the balance error scoring system. *Sports Health, 3*(3), 287–295.
7. Biz, C., Nicoletti, P., Tomasin, M., Bragazzi, N. L., Di Rubbo, G., & Ruggieri, P., (2022). Is Kinesio taping effective for sports performance and ankle function of athletes with chronic ankle instability (CAI)? A systematic review and meta-analysis. *Medicina, 58*(5), 620.
8. Braun, G. L., (1941). Kinesiology: From Aristotle to the twentieth century. *Research Quarterly. American Association for Health, Physical Education and Recreation, 12*(2), 163–173.
9. Carini, F., Mazzola, M., Fici, C., Palmeri, S., Messina, M., Damiani, P., & Tomasello, G., (2017). Posture and post urology, anatomical and physiological profiles: Overview and current state of art. *Acta Bio Medica: Atenei Parmensis, 88*(1), 11.
10. Chi, L., & Kennon, R., (2006). Body scanning of dynamic posture. *International Journal of Clothing Science and Technology*.
11. Clauser, C. E., McConville, J. T., & Young, J. W., (1969). *Weight, Volume, and Center of Mass of Segments of the Human Body (pp 1-25)*.

12. Conable, K. M., (2010). Intraexaminer comparison of applied Kinesiology manual muscle testing of varying durations: A pilot study. *Journal of Chiropractic Medicine, 9*(1), 3–10.
13. Cuccia, A., & Caradonna, C., (2009). The relationship between the stomatognathic system and body posture. *Clinics, 64*(1), 61–66.
14. Dempster, W. T., & Gaughran, G. R., (1967). Properties of body segments based on size and weight. *American Journal of Anatomy, 120*(1), 33–54.
15. Dishman, R. K., Washburn, R. A., & Schoeller, D. A., (2001). Measurement of physical activity. *Quest, 53*(3), 295–309.
16. Donald, A., Shimada, T., & Hirata, S., (trans.)(2006). *Kinesiology of the Musculoskeletal System(pp.1-5).*
17. Elvan, A., & Ozyurek, S., (2020). Principles of Kinesiology. In: *Comparative Kinesiology of the Human Body* (pp. 13–27). Elsevier.
18. Enoka, R. M., (1988). *Neuromechanical Basis of Kinesiology*: ERIC.
19. Frost, R., (2013). *Applied Kinesiology (Revised Edition): A Training Manual and Reference Book of Basic Principles and Practices*. North Atlantic Books.
20. Gin, R. H., & Green, B. N., (1997). George Goodheart, Jr., DC, and a history of applied Kinesiology. *Journal of Manipulative and Physiological Therapeutics, 20*(5), 331–337.
21. Goodheart, B., & Cox, J., (1994). *The Magic Garden Explained: The Internals of UNIX System V Release 4: An Open Systems Design*: Prentice-Hall, Inc.
22. Goodheart, C. D., Kazdin, A. E., & Sternberg, R. J., (2006). *Evidence-Based Psychotherapy: Where Practice and Research Meet*: American Psychological Association.
23. Gribble, P. A., Delahunt, E., Bleakley, C., Caulfield, B., Docherty, C., Fourchet, F., & Kaminski, T., (2013). Selection criteria for patients with chronic ankle instability in controlled research: A position statement of the international ankle consortium. *Journal of Orthopedic & Sports Physical Therapy, 43*(8), 585–591.
24. Hall, S., Lewith, G., Brien, S., & Little, P., (2008). A review of the literature in applied and specialized Kinesiology. *Complementary Medicine Research, 15*(1), 40–46.

25. Hamill, J., Knutzen, K. M., & Derrick, T. R., (2021). Biomechanics: 40 years on. *Kinesiology Review, 10*(3), 228–237.
26. Hamilton, C., Miller, A., Casablanca, Y., Horowitz, N., Rungruang, B., Krivak, T., & Backes, F., (2018). Clinicopathologic characteristics associated with long-term survival in advanced epithelial ovarian cancer: An NRG oncology/gynecologic oncology group ancillary data study. *Gynecologic Oncology, 148*(2), 275–280.
27. Heggers, J., Goodheart, R. E., Washington, J., McCoy, L., Carino, E., Dang, T., & Chinkes, D., (2005). The Lindberg award: Therapeutic efficacy of three silver dressings in an infected animal model. *The Journal of Burn Care & Rehabilitation, 26*(1), 53–56.
28. Jackson, K., Simon, J. E., & Docherty, C. L., (2016). Extended use of Kinesiology tape and balance in participants with chronic ankle instability. *Journal of Athletic Training, 51*(1), 16–21.
29. Jacobs, G., Franks, T., & Gilman, P., (1984). Diagnosis of thyroid dysfunction: Applied Kinesiology compared to clinical observations and laboratory tests. *Journal of Manipulative and Physiological Therapeutics, 7*(2), 99–104.
30. Johansson, R. S., Backlin, J. L., & Burstedt, M. K., (1999). Control of grasp stability during pronation and supination movements. *Experimental Brain Research, 128*(1), 20–30.
31. Jonsson, B., (1978). Kinesiology: With special reference to electromyographic Kinesiology. *Electroencephalography and Clinical Neurophysiology. Supplement*, *34*, 417–428.
32. Kage, C. C., Akbari-Shandiz, M., Foltz, M. H., Lawrence, R. L., Brandon, T. L., Helwig, N. E., & Ellingson, A. M., (2020). Validation of an automated shape-matching algorithm for biplane radiographic spine osteokinematics and radiostereometric analysis error quantification. *Plos One, 15*(2), e0228594.
33. Kapandji, A., (2001). Biomechanics of pronation and supination of the forearm. *Hand Clinics, 17*(1), 111–122, VII.
34. Kaptchuk, T. J., & Eisenberg, D. M., (1998). Chiropractic: Origins, controversies, and contributions. *Archives of Internal Medicine, 158*(20), 2215–2224.
35. Ko, K. J(2000). *Osteokinematics and Arthrokinematics of the Hip.*(pp. 7-9)

36. Lafon, Y., Smith, F. W., & Beillas, P., (2010). Combination of a model-deformation method and a positional MRI to quantify the effects of posture on the anatomical structures of the trunk. *Journal of Biomechanics, 43*(7), 1269–1278.
37. Leach, R. A., (2004). *The Chiropractic Theories: A Synopsis of Scientific Research.*(Vol. 1, pp. 1-6)
38. Lee, D., (1993). Biomechanics of the thorax: A clinical mode of *in vivo* function. *Journal of Manual & Manipulative Therapy, 1*(1), 13–21.
39. Lee, S. H., Kim, Y., Lee, D. G., Lee, K. B., & Lee, G. C., (2017). Osteokinematic analysis during shoulder abduction using the C-arm. *Physical Therapy Rehabilitation Science, 6*(4), 208–213.
40. Lee, S. M., & Lee, J. H., (2017). The immediate effects of ankle balance taping with Kinesiology tape on ankle active range of motion and performance in the balance error scoring system. *Physical Therapy in Sport, 25*, 99–105.
41. Lichtenstein, E., Donath, L., Oeri, A., Roth, R., Rössler, R., Zahner, L., & Faude, O., (2018). Performance, stride characteristics, and muscle activity while running with a traditional compared to a newly developed running shoe. *Kinesiology, 50*(1), 126–132.
42. Lobbezoo, F., Van, D. Z. J., Visscher, C., & Naeije, M., (2004). Oral Kinesiology. A new postgraduate program in the Netherlands. *Journal of Oral Rehabilitation, 31*(3), 192–198.
43. Loper, M., Mahmood, N., & Black, M. J., (2014). MoSh: Motion and shape capture from sparse markers. *ACM Transactions on Graphics (ToG), 33*(6), 1–13.
44. Lowrie, P. M., & Robinson, L. E., (2013). Creating an inclusive culture and climate that supports excellence in Kinesiology. *Kinesiology Review, 2*(3), 170–180.
45. Lutgendorf, S. K., DeGeest, K., Dahmoush, L., Farley, D., Penedo, F., Bender, D., & Krueger, G., (2011). Social isolation is associated with elevated tumor norepinephrine in ovarian carcinoma patients. *Brain, Behavior, and Immunity, 25*(2), 250–255.
46. Maugeri, G., D'Agata, V., Roggio, F., Cortis, C., Fusco, A., Foster, C., & Piacentini, M. F., (2020). *The "Journal of Functional Morphology and Kinesiology" Journal Club Series: PhysioMechanics of Human Locomotion* (Vol. 5, pp. 52). Multidisciplinary Digital Publishing Institute.

47. McAfee, M., (2014). The Kinesiology of race. *Harvard Educational Review, 84*(4), 468–491.

48. McDonnell, R., Jörg, S., Hodgins, J. K., Newell, F., & O'sullivan, C., (2009). Evaluating the effect of motion and body shape on the perceived sex of virtual characters. *ACM Transactions on Applied Perception (TAP), 5*(4), 1–14.

49. Neuman, D. A., (2010). *Kinesiology of the Musculoskeletal System* (pp. 3–26). Seoul: Jungdam Media.

50. Neumann 3rd, D., (2017). *Kinesiology of the Musculoskeletal System: Foundations for Rehabilitation [Google Scholar]*. Elsevier; St Louis.

51. Neumann, D. A., (2010a). Kinesiology of the hip: A focus on muscular actions. *Journal of Orthopedic & Sports Physical Therapy, 40*(2), 82–94.

52. Neumann, D. A., (2010b). *Kinesiology of the Musculoskeletal System; Foundation for Rehabilitation.* Mosby & Elsevier.

53. Neumann, D. A., (2016). *Kinesiology of the Musculoskeletal System-E-Book: Foundations for Rehabilitation*: Elsevier Health Sciences.

54. Neumann, D., (2010). *Kinesiology of the Musculoskeletal System: Foundations for Rehabilitation.* Mosby, Inc., an affiliate of Elsevier Inc.

55. Nigg, B. M., MacIntosh, B. R., & Mester, J., (2000). *Biomechanics and Biology of Movement*: Human Kinetics.

56. Nirendan, J., & Murugavel, K., (2019). *Impact of Low Intensity Sports Specific Resistance Training with Yoga on Selected Osteokinematics Variables of Badminton Players*.(Vol. 1, pp. 7-12)

57. Paulovich, J. M., (2018). *Kinesiology Tape and its Effects on Postural Control.* Ohio University.

58. Perry, J., (1974). Kinesiology of lower extremity bracing. *Clinical Orthopedics and Related Research (1976–2007), 102*, 18–31.

59. Rosheim, M. E., (1997). In the footsteps of Leonardo~ articulated anthropomorphic robot. *IEEE Robotics & Automation Magazine, 4*(2), 12–14.

60. Salvia, P., Woestyn, L., David, J. H., Feipel, V., Van, S., Jan, S., & Rooze, M., (2000). Analysis of helical axes, pivot and envelope in active wrist circumduction. *Clinical Biomechanics, 15*(2), 103–111.

61. Sherman, D., Fuller, P. M., Marcus, J., Yu, J., Zhang, P., Chamberlin, N. L., & Lu, J., (2015). Anatomical location of the mesencephalic locomotor region and its possible role in locomotion, posture, cataplexy, and parkinsonism. *Frontiers in Neurology, 6*, 140.
62. Shih, Y. F., Lee, Y. F., & Chen, W. Y., (2018). Effects of Kinesiology taping on scapular reposition accuracy, kinematics, and muscle activity in athletes with shoulder impingement syndrome: A randomized controlled study. *Journal of Sport Rehabilitation, 27*(6), 560–569.
63. Shorter, K. A., Wu, A., & Kuo, A. D., (2017). The high cost of swing leg circumduction during human walking. *Gait & Posture, 54*, 265–270.
64. Simeón, F., & Monge, J. C., (2005). Kinesiology. *Revista de Enfermeria (Barcelona, Spain), 28*(12), 19–22.
65. Slaughter, V., Heron, M., & Sim, S., (2002). Development of preferences for the human body shape in infancy. *Cognition, 85*(3), B71–B81.
66. Sloane, R. B., (1952). Kinesiology and vertical dimension. *The Journal of Prosthetic Dentistry, 2*(1), 12–14.
67. Smith, M. D., Vitharana, T. N., Wallis, G. M., & Vicenzino, B., (2020). Response profile of fibular repositioning tape on ankle osteokinematics, arthrokinematics, perceived stability and confidence in chronic ankle instability. *Musculoskeletal Science and Practice, 50*, 102272.
68. Swanson, S. C., & Frappier, J. P., (2001). Interdisciplinary tool. *Sports Injury: Prevention & Rehabilitation*, 1.
69. Thompson, D. M., (2003). Kinesiology of the musculoskeletal system: Foundations for physical rehabilitation. *Physical Therapy, 83*(4), 402.
70. Tomchuk, D., (2010). *Companion Guide to Measurement and Evaluation for Kinesiology*. Jones & Bartlett Publishers.
71. Twietmeyer, G., (2012). What is Kinesiology? Historical and philosophical insights. *Quest, 64*(1), 4–23.
72. Vouloutsi, V., Halloy, J., Mura, A., Mangan, M., Lepora, N., Prescott, T. J., & Verschure, P. F., (2018). *Biomimetic and Biohybrid Systems: 7th International Conference, Living Machines 2018, Paris, France, Proceedings* (Vol. 10928). Springer.
73. Yin, L., & Wang, L., (2020). Acute effect of Kinesiology taping on postural stability in individuals with unilateral chronic ankle instability. *Frontiers in Physiology, 11*, 192.

74. Zampini, M., Harris, C., & Spence, C., (2005). Effect of posture change on tactile perception: Impaired direction discrimination performance with interleaved fingers. *Experimental Brain Research, 166*(3), 498–508.

CHAPTER 2

KINESIOLOGY OF SHOULDER COMPLEX

CONTENTS

2.1. Introduction 34

2.2. Structure of the Bones of the Shoulder Complex 34

2.3. The Shoulder Complex's Joint Structure and Supporting Structures 47

2.4. Muscle Activity in the Shoulder Complex: Pathomechanics and Mechanics 57

2.5. The Relationship Between Joint and Muscle Force Analysis and Clinical Practice 66

References 67

2.1. INTRODUCTION

The shoulder complex is the functional unit responsible for arm mobility with the trunk. This unit comprises the humerus, scapula, and clavicle as well as their connecting articulations and the muscles which move them. These structures are so operationally interdependent that it is nearly difficult to analyze their independent roles. A thorough examination of the components that make up the shoulder unit, though, shows an exquisitely basic system of joints, muscles, and bones that permit the shoulder to perform an unlimited number of motions. A disruption in the normal coordination of such interdependent structures is a significant cause of patients' complaints of disability and pain in the shoulder complex (Bigliani et al., 1996; Branch et al., 1995; Curl and Warren, 1996; Hislop and Montgomery, 1995).

The major purpose of the shoulder complex is to arrange the upper limb in space so that the hand may accomplish its functions. The miracle of the shoulder complex is the range of postures it may acquire; however, such mobility also poses a significant threat to the shoulder complex. The instability of the joint is another significant cause of shoulder dysfunction symptoms among patients. Understanding the disorder and function of the shoulder complex necessitates gratitude for the coordinated interaction between the individual elements of the shoulder complex and the structural complication observed in the shoulder which permits great mobility while still providing adequate stability (Deutsch et al., 1996; Ebaugh, McClure, and Karduna, 2005; Field et al., 1995; Flatow et al., 1995; Graichen et al., 2000).

2.2. STRUCTURE OF THE BONES OF THE SHOULDER COMPLEX

The humerus, scapula, and clavicle are the three bones that make up the shoulder complex. All of these bones are explored in further depth below. The complex, on either hand, is related to the axioskeleton through the sternum and lies on the thorax; its structure influences the complex's function. As a result, a brief explanation of the thorax morphology and sternum with the shoulder complex has been presented (Graichen et al., 2000; Harrington et al., 1993; Harryman 2nd et al., 1990).

2.2.1. Clavicle

The clavicle acts as a strut to hang the shoulder complex and the whole top extremity from the axioskeleton (GJE, 1981). Other roles of the clavicle include providing a place for the attachment of muscles, protecting underlying blood vessels and nerves, contributing to enhanced shoulder range of motion, and aiding in muscle force transmission to the scapula (Kent, 1971; Moseley, 1968). This section explains the clavicular characteristics that contribute to its capacity to fulfill all of these activities.

According to the previous knowledge, we will look at how such traits influence the clavicle's functions and how they are linked to clavicle injuries. The long axis of the clavicle is near the transverse plane. When observed upwards side, it has a crank-shaped appearance, with the medial 2/3 convex interiorly, roughly corresponding to the anterior thorax and the lateral 1/3 convex posteriorly (Figure 2.1). When discussing total shoulder mobility, the functional relevance of such an odd design becomes clear (Harryman, Sidles, Harris, and Matsen, 1992; Hjelm, Draper, and Spencer, 1996; Howell and Galinat, 1989; Itoi et al., 1998).

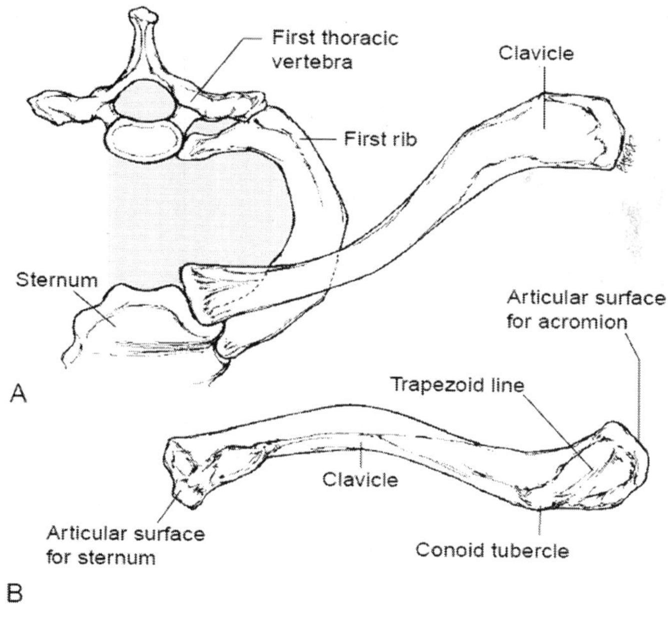

Figure 2.1. Clavicle. (A) View of the superior surface; (B) view of inferior surface.

Source: https://boneandspine.com/anatomy-clavicle/.

The clavicle's superior surface is smooth and easily palpable under the skin. The connections of the pectoralis main medially and the deltoid laterally roughen the surface anteriorly. The connection of the upper trapezius roughens the posterior surface on the lateral 1/3. The costoclavicular ligament and the subclavius muscle adhere medially to the surface, while the coracoclavicular ligament attaches laterally (Bigliani et al., 1996; Curl and Warren, 1996; Graichen et al., 2000). The conoid tubercle and, lateral to it, the trapezoid line, are two conspicuous marks on the inferior side of the lateral side of the clavicle.

The acromion and sternum have articular surfaces on the lateral and medial extremities of the clavicle, correspondingly. The clavicle's medial portion extends to produce the clavicle's head. The sternum and intervening articular disc, or meniscus, and 1st costal cartilage, interface with the medial surface of such an extension. The clavicular head's articular surface is somewhat convex in the superior-inferior (SI) axis and concaves in the anterior-posterior (AP) axis (Stokdijk, Eilers, Nagels, and Rozing, 2003; Williams et al., 1995). Unlike other synovial joints, the adult clavicle's articular surface is coated with dense fibrocartilage. The clavicle's latter 3rd is flatter than the first 2/3, ending in a wide flat expansion that articulates with the acromion at the acromioclavicular joint. A tiny facet on the articular surface usually faces laterally and inferiorly. Instead of hyaline cartilage, it is also coated by fibrocartilage. The clavicle's lateral and medial facets are readily palpable (Jobe, 1997; Jobe and Lannotti, 1995; Karduna, Williams, and Iannotti, 1997).

2.2.2. Scapula

The scapula is a flat bone with the primary purpose of providing a location for the shoulder's muscle attachment. The scapula is the attachment point for 15 main shoulder muscles (Lucas, 1973; Williams et al., 1995). The scapula of quadrupedal animals is slender and long, resting on the lateral portion of the thorax. There is a steady mediolateral growth of the bone in primates, as well as a gradual movement from a lateral towards a more posterior point on the thorax (Figure 2.2). The expanded infraspinous costal and fossa surface, which provide attachment for 3/4 rotator cuff muscles and also numerous additional shoulder muscles, contribute to the mediolateral enlargement (Inman, Saunders, and Abbott, 1944; Roberts, 1974). The progressive shift in the upper extremity's role from weight-bearing to grasping and reaching is reflected in such changes in scapula form and position (Ben, 1998; Kennedy and Cameron, 1954; Kuhn et al., 2005). These changes in function necessitate a shift in muscle function, as the scapula and glenohumeral joint should now position and support a scapula and glenohumeral joint that are no longer predominantly weight-bearing and may now move over a considerably greater range of motion (Levinsohn and Santelli, 1991; Lewis, Ballet, Kroll, and Bloom, 1985; Magermans et al., 2005).

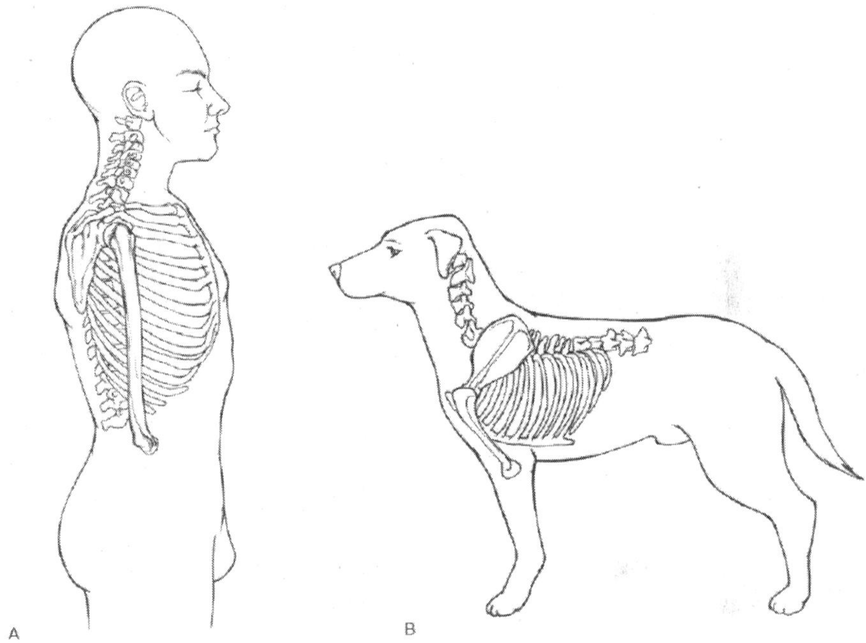

Figure 2.2. Location of the scapula. (A) The scapula has been more posteriorly placed in humans. (B) In quadrupedal animals, the scapula is positioned on the lateral portion of the thorax.

Source: http://pressbooks-dev.oer.hawaii.edu/anatomyandphysiology/chapter/the-pectoral-girdle/.

The scapula comprises two surfaces, the dorsal or posterior surface and the costal or anterior surface (Figure 2.3). The costal surface is normally smooth and serves as the muscle's proximal attachment point. A smooth, thin surface gives rise to the serratus anterior muscle along the medial border of the anterior surface (McClure, Michener, Sennett, and Karduna, 2001; McQuade and Smidt, 1998; Michener, McClure, and Karduna, 2003).

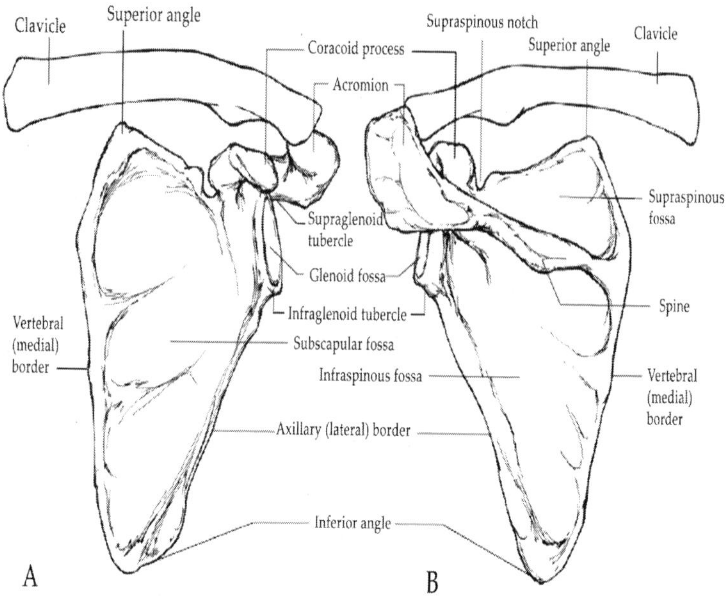

Figure 2.3. Scapula. (A) Anterior surface; (B) posterior surface.

Source: https://musculoskeletalkey.com/fractures-of-the-scapula/.

The spine of the scapula divides the dorsal side of the scapula into two areas: a little superior area is known as the supraspinous fossa and a big inferior space is known as the infraspinous fossa. The spine is a broad dorsally projecting ridge of bone that extends superiorly and laterally across the breadth of the scapula from the medial border. The spine terminates in a broad, flat surface that protrudes anteriorly, laterally, and somewhat superiorly. The acromion procedure is the name for this method. The acromion serves as a roof over the humerus' head. On the anterior portion of its medial surface, the acromion possesses an articular face for the clavicle (Moseley and Övergaard, 1962; Nettles and Linscheid, 1968; Neviaser, 1987).

The articular surface, such as the clavicular surface in which it articulates, is coated in fibrocartilage instead of hyaline cartilage. This aspect is oriented medially and slight superiorly. The acromion is often defined as being flat. Bigliani et al. on either hand, characterize the acromion as having rounded, flat, and hooked procedures (Bigliani, 1986). The hooked acromion procedures, according to these scientists, can lead to shoulder impingement disorders. This chapter goes over other elements that contribute to impingement syndrome (Neumann, Petersen, and Jahnke, 1991; Norkin and White, 2016; O'Connell et al., 1990; Pagnani et al., 1995; Pratt, 1994; Soslowsky, Malicky, and Blasier, 1997).

There are three borders on the scapula: the superior border, the axillary or lateral border, and the vertebral or medial border. The medial boundary is palpable from inferior to superior over its entire length. From the root of the spine to the superior angle, the medial border bends anteriorly, adhering to the curves of the underlying thorax. It connects to the superior border of the scapula at the superior angle, which may only be felt in those who have tiny or atrophied muscles covering the superior angle, notably the elevator and trapezius scapulae (Spencer et al., 2005; Steindler, 1977; Terry, Hammon, France, and Norwood, 1991; Thomas and Friedman, 1989).

The coracoid procedure is a finger-like protrusion that protrudes laterally, anteriorly, and superiorly from the scapula and projects from the anterior surface of the superior scapular border. It is roughly 2/3 of the scapula's breadth from its medial edge. On the anterior part of the trunk, the coracoid process is easily palpable inferior to the lateral 1/3 of the clavicle. The supraspinous notch, located just medial to the base of the coracoid process on the superior border, is where the suprascapular nerve passes (Figure 2.4) (Wuelker, Korell, and Thren, 1998; Yanai, Fuss, and Fukunaga, 2006).

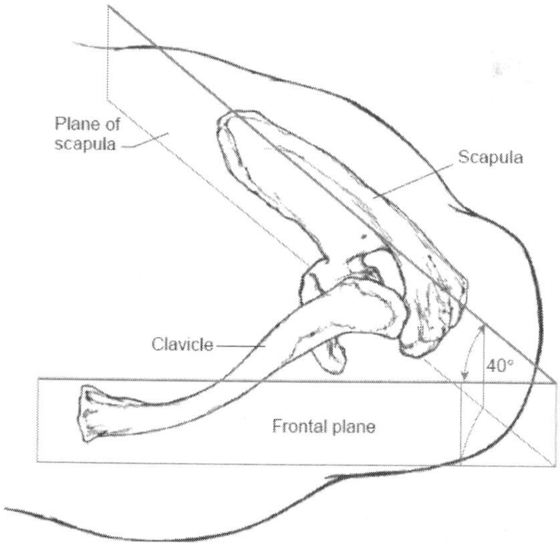

Figure 2.4. Plane of the scapula. The plane of the scapula makes an angle of roughly 40° with the frontal plane when seen transversely.

Source: https://quizlet.com/337327276/bk-normal-shoulder-flash-cards/.

At the inferior angle, the scapula's medial border meets the lateral border, forming an effective and easily recognized feature. The scapula's lateral border is perceptible along its inferior section till the teres major, teres minor, and latissimus dorsi muscles surround it. At the anterior angle or neck and head of the scapula, the lateral border extends superiorly and joins the superior border. The glenoid fossa is formed by the head and gives the glenohumeral joint with an articular surface on the scapula. The fossa has a "pear-shaped" look because it is small superiorly and expands inferiorly. The adjacent fibrocartilaginous labrum adds to the depth of the fossa. The infraglenoid and supraglenoid tubercles, correspondingly, are above and inferior to the fossa (Figure 2.5).

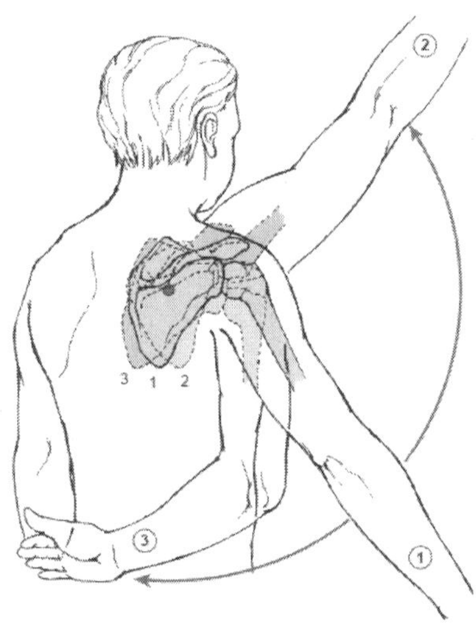

Figure 2.5. Scapular rotation. Rotation of the scapula about an anterior–posterior (AP) axis causes the glenoid fossa to face upward (2) or downward (3).

Source: https://www.researchgate.net/figure/Lateral-upward-rotation-of-scapular-motion-during-90-8-anterior-flexion-of-the_fig1_280999234.

The direction of the glenoid fossa is a point of contention. It has the following orientation:

- Lateral (Jv and De Luca, 1985);
- Superior (Jv and De Luca, 1985);
- Anterior (Jv and De Luca, 1985; Roberts, 1974);
- Inferior (Nk, 1976);
- Retroverted (Saha, 1973).

Only the glenoid fossa's lateral alignment seems to be unchallenged. While inconsistencies in the literature can represent genuine disparities in measurement or populations investigated, minimum of a few of the variation is attributable to changes in the reference frames utilized by different investigators to define the location of the scapula. One reference frame is embedded in the scapula itself, whereas the other is embedded in the entire body. The scapula-fixed reference frame permits you to compare the location of one scapula bone marker to another. The latter body-fixed reference frame facilitates the comparison of scapular landmark location to other body areas.

To comprehend the issues surrounding the direction of the glenoid fossa, it is also important to 1st evaluate the overall alignment of the scapula. The natural resting posture of the scapula may be characterized relative to the sagittal, frontal, and transverse planes utilizing a body-fixed reference frame. The scapula is turned inward about a vertical axis when seen in a transverse plane. Approximately 30° to 50° separate the plane of the scapula from the frontal plane (Figure 2.4) (Kapandji, 1982; Saha, Das, and Dutta, 1983). This posture positions the glenoid anteriorly relative to the body. A scapula-fixed reference frame, although, indicates that the glenoid fossa is retroverted, or turned posteriorly, relative to the scapula's neck (Figure 2.6) (Couteau et al., 2000; Saha, 1973).

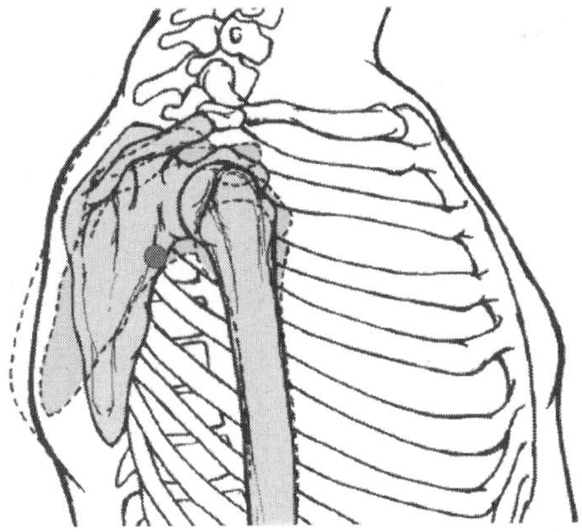

Figure 2.6. Scapular rotation. The scapula tilts anteriorly and posteriorly when rotated about an ML axis.

Source: https://www.researchgate.net/figure/Individual-axes-and-rotations-used-to-describe-scapular-orientation-and-position-A_fig1_24209934.

Consequently, the glenoid fossa has been oriented anteriorly (relative to the body) and retroverted (relative to the scapula).

The scapula is also rotated in the frontal plane around a body-fixed AP axis (Figure 2.5). The glenoid fossa's downward or upward orientation, or the lateral or medial placement of the scapula's inferior angle, explains this frontal plane rotation of the scapula (Freedman and Munro, 1966; Jv and De Luca, 1985; Nk, 1976). A rotation along the AP axis that tips the glenoid fossa inferiorly and shifts the inferior angle of the scapula medially is characterized as a medial or downward rotation of the scapula. This can also be referred to as a downward rotation of the scapula (for instance, nearer to the vertebral column). Lateral or u Upward rotation is a rotation which tilts the glenoid fossa upward, shifting the inferior angle laterally far from the spinal column. In silent standing, the glenoid fossa is upwardly inclined, according to two studies (Jv and De Luca, 1985; Lukasiewicz et al., 1999). Two more investigations (Freedman and Munro, 1966; Nk, 1976) found a downward inclination of around 5°.

The individual's posture in the experiments might explain the apparent variances. Those participants who have a glenoid fossa that slopes upward can be instructed to bring their shoulders back up into an "erect" position, whereas those participants whose glenoid fossa slopes downward have shoulders that are somewhat stooped (Figure 2.6). To provide a conclusive assessment of the normal alignment of the scapulae in the frontal plane, it is necessary to have a well-defined description of the typical postural orientation of the shoulder. However, it is the current definition.

2.2.3. Proximal Humerus

The humerus is a long bone with a neck, head, and body, also known as the shaft (Figure 2.7). The capitulum and trochlea are the distal extremities of the body. Only the parts of the humerus that are important to the pathomechanics and mechanics of the shoulder complex are covered in this section. The elbow and the remainder of the humerus are also described in the following chapters. The articular surface of the humeral head is commonly characterized as 50% of the perfect sphere (Figure 2.8) (Iannotti et al., 1992; Soslowsky et al., 1992; van der Helm and Pronk, 1995; Williams et al., 1995). With the plane created by the lateral and medial condyles, the humeral head extends superiorly, medially, and posteriorly (Figure 2.9) (Inokuchi, Olsen, Søjbjerg, and Sneppen, 1997). The articular surface of the humeral head terminates at the anatomical neck.

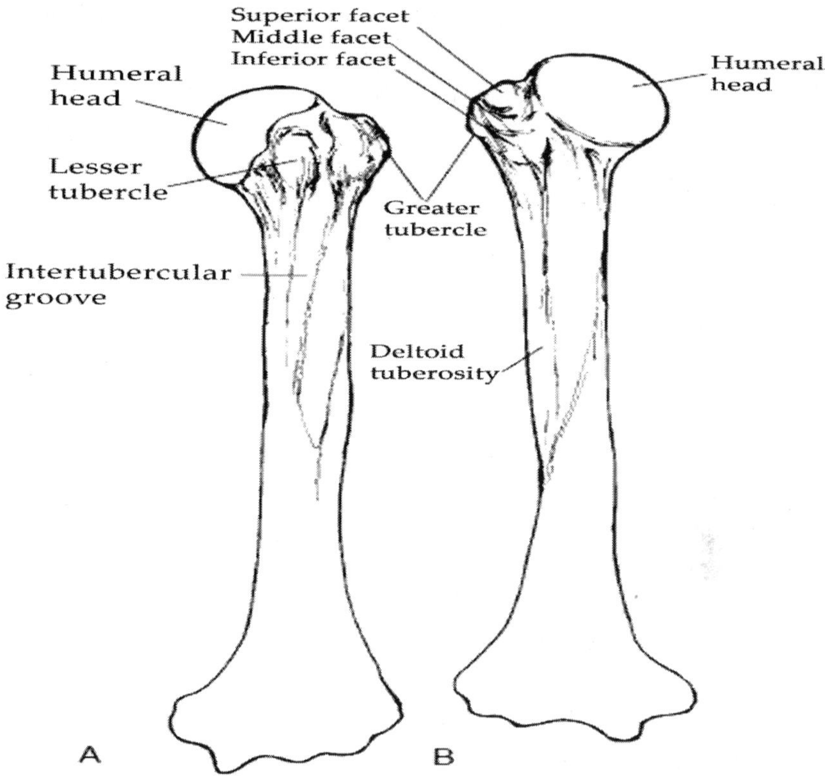

Figure 2.7. Proximal humerus. (A) Anterior view; (B) posterior view.

Source: https://doctorlib.info/medical/anatomy/21.html.

The larger tubercle is a prominent bony protrusion on the lateral side of the proximal humerus that has been readily palpated on the lateral side of the shoulder complex. The posterior and superior surfaces of the larger tubercle have three different aspects. From superior to posterior, these aspects give way to the infraspinatus, teres minor muscles, and supraspinatus. The lesser tubercle is a small but still conspicuous bony protrusion on the front part of the proximal humerus. It has a facet that connects to the subscapularis, the residual rotator cuff muscle.

The bicipital, or intertubercular, groove separates the tubercles and contains the tendon of the long head of the biceps brachii. As the lateral and medial lips of the groove, the larger and lesser tubercles extend onto the body of the humerus. A little narrowing of the shaft of the humerus immediately distal to the tubercles is called the surgical neck.

On the anterolateral surface of the humerus, the deltoid tuberosity is located about halfway distally on the body. It serves as the deltoid muscle's distal connection. The spiral groove on the humerus's body seems to be another essential marker. On the posterior side, it spirals from proximal to distal and medial to lateral on the proximal half of the humerus. Along with the profound brachii vessels, the radial nerve passes along the spiral groove. Because it sits in the spiral groove, the radial nerve is especially vulnerable to damage.

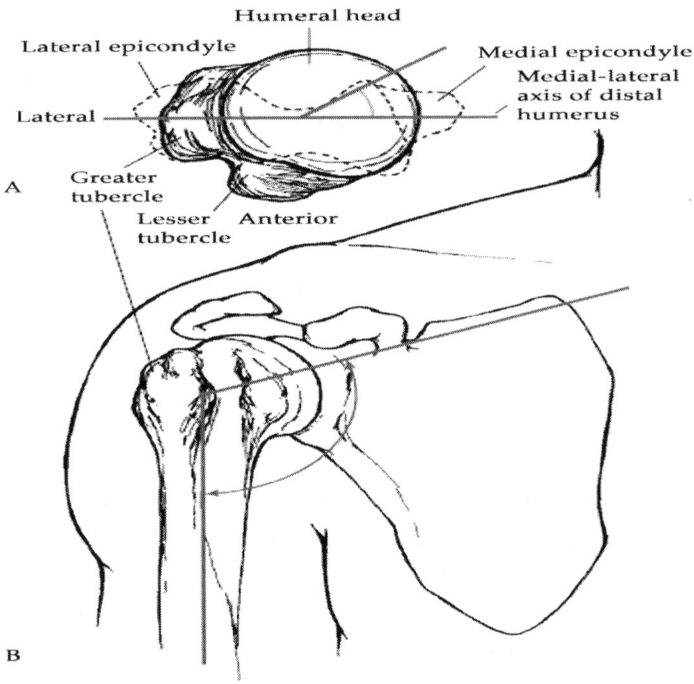

Figure 2.8. Orientation of the humerus's head. (A) The humeral head is rotated posteriorly in the transverse plane with the distal humeral condyles. (B) In the frontal plane, the humeral head is inclined superiorly and medially about the humeral shaft.

Source: https://www.physio-pedia.com/Hip_Anatomy.

2.2.4. Thorax and Sternum

The thorax and sternum are not components of the shoulder complex, but they are intimately associated with it, thus a brief discussion of their construction with the shoulder complex is necessary. The manubrium, the superior region of the sternum, offers an articular surface for every clavicle's proximal part (Figure 2.9). The shallow depression is called the clavicular notch and that covers the articular surface, such as the clavicular head in which it articulates. Fibrocartilage has been covered over it. Every notch has an articular surface that is smaller than the surface of the clavicular head with which it articulates. The jugular or sternal notch on the superior part of the manubrium separates the two clavicular notches. This noticeable notch serves as a handy signpost for locating the sternoclavicular joints. The angle of Louis, or sternal angle, is another dependable and useful diagnostic mark generated by the intersection of the manubrium with the body of the sternum. The 2nd costal cartilage is attached to the manubrium and the body of the sternum at this location.

The skeletal thorax serves as the base onto whereby the two scapulae glide. Therefore, the structure of the thorax limits the mobility of the scapulae (Van der Helm, 1994). Every scapula rest on the upper section of the thorax, which in an upright position extends about from the 1st to the 18 ribs and the T2 to T7, or T8 vertebral bodies. The medial side of the scapular spine is commonly characterized as being parallel to the spinous procedure of T2.

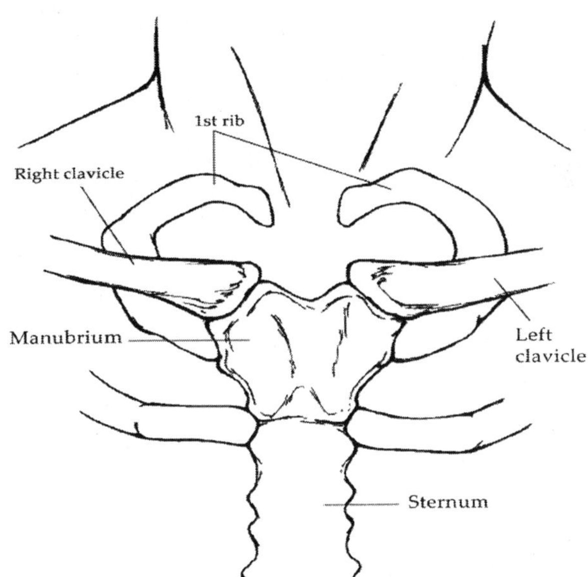

Figure 2.9. This is the articular surface of the sternum. The sternum supplies the clavicle's head with a shallow articular surface.

Source: https://www.earthslab.com/anatomy/sternoclavicular-joint/.

The inferior angle is commonly said to be parallel to T7's spinous procedure. It is vital to remember, nevertheless, that the shoulder and spinal column's postural alignment may drastically modify such connections. The convex form of the dorsal surface of the thorax in the area of the scapulae is termed thoracic kyphosis. Because the superior ribs are smaller than the inferior ones, the thorax's overall form is elliptical (Figure 2.10) (Van der Helm et al., 1992). The scapula tilts anteriorly as it slides superiorly on the thorax. The resting position of the scapula and the movements of the scapula generated by contractions of particular muscles like the rhomboids and pectoralis minor may be explained by understanding the form of the thorax upon which the scapula slides (Culham and Peat, 1993; Kebaetse, McClure, and Pratt, 1999).

To summarize, the shoulder complex is an elaborate organization of three main bones, and each one is unique. Components of the axioskeleton are physically and functionally connected to these three bones (for example, to the thorax and the sternum). A full and accurate physical examination requires a clear image of every bone and its location concerning the others.

The following are important bone landmarks in the shoulder complex:

- Sternal angle;
- Sternal notch;
- Head of the clavicle;
- Second rib;
- The superior surface of the clavicle;
- Sternoclavicular joint;
- The anterior surface of the clavicle;
- Acromioclavicular joint;
- Acromion;
- Vertebral border of the scapula;
- Coracoid process;
- The intertubercular groove of the humerus;
- The inferior angle of the scapula;
- The spine of the scapula;
- Axillary border of the scapula;
- Lesser tubercle of the humerus;
- Greater tubercle of the humerus.

The anatomy and mechanics of the joints of the shoulder complex created by such bony components are described in the next section.

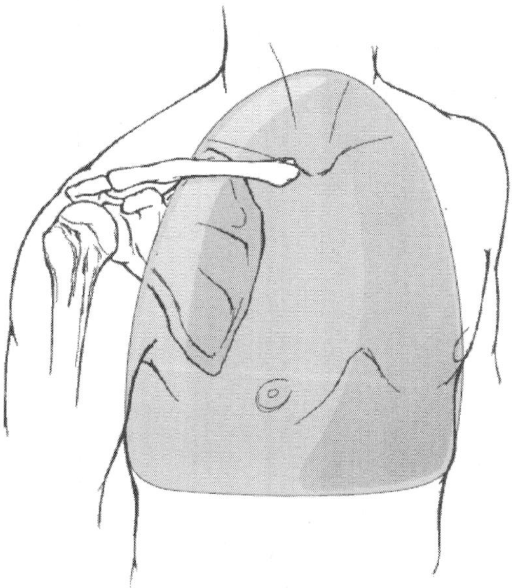

Figure 2.10. The shape of the thorax. The scapula's mobility is influenced by the elliptical form of the thorax.

Source: https://theodora.com/anatomy/the_thorax.html.

2.3. THE SHOULDER COMPLEX'S JOINT STRUCTURE AND SUPPORTING STRUCTURES

The shoulder complex consists of the following four joints:

- Acromioclavicular;
- Sternoclavicular;
- Glenohumeral; and
- Scapulothoracic.

Except for the scapulothoracic joint, all of the joints are synovial. Since the moving elements, the thorax, and the scapula, are not directly linked or articulated to each other, and because muscles, not cartilage or fibrous material, divide the moving elements, the scapulothoracic joint falls outside of any traditional categorization of joint. Therefore, because it is the site of regular and recurrent motion among bones, it may legitimately be called a joint. The construction and mechanics of every one of the four joints of the shoulder complex are described as follows.

2.3.1. Sternoclavicular Joint

The lateral, posterior, and anterior motions are all limited by the capsule and ligaments discussed thus far. Other structures, although, place further restrictions on clavicle elevation and medial translation. The articular surface of the clavicle is significantly bigger than the corresponding region on the sternum, as stated in the bone classifications (Figure 2.11).

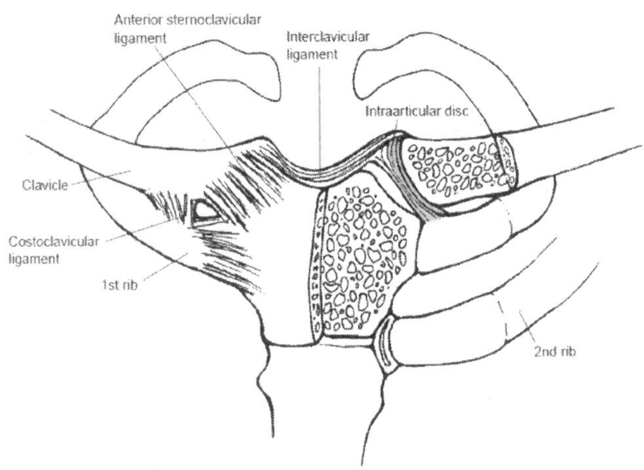

Figure 2.11. The sternoclavicular joint. The intraarticular disc, the capsule, the interclavicular ligament, the posterior and anterior sternoclavicular ligaments, and the costoclavicular ligament are the supporting components of the sterno-clavicular joint.

Source: https://www.kenhub.com/en/library/anatomy/sternoclavicular-joint.

As a result, the superior part of the clavicular head protrudes above the sternum and can be felt. Because of the difference in articular surfaces, the clavicle can slip medially across the sternum, causing joint instabilities. A medically directed force on the clavicle, like that caused by a blow to or a fall on the shoulder, may cause this migration (Figure 2.12). Interposed between the sternum and the clavicle, and intraarticular disc enhances the articular surface upon which the clavicle travels while also blocking any medial motion. The disc connects to the superior side of the 1st costal cartilage and the superior border of the clavicle's articular surface, splitting the joint into two synovial chambers. The clavicle's medial migration over the sternum is prevented by the disc's particular attachments.

The clavicle is pushed medially on the sternum by a blow to the lateral side of the shoulder. The intraarticular disc anchors the clavicle to the underlying 1st costal cartilage, preventing it from moving medially. A cadaver investigation, however, reveals that the disc may be readily ripped from its connection to the costal cartilage (Bearn, 1967). As a result, the degree of its involvement as a medial clavicle translation on the sternum is unknown. Between the sternum and the clavicle, the disc may act as a stress absorber (Kelley and Clark, 1995).

A further significant supporting element of the sternoclavicular joint is the costoclavicular ligament, a lateral, extracapsular ligament. It connects the superior lateral side of the 1st costal cartilage to the inferior aspect of the medial clavicle. The anterior fibers of this muscle run laterally and superiorly, whereas the posterior fibers travel medially and superiorly. Therefore, this ligament significantly restricts lateral, medial, posterior, and anterior clavicular motions, and also elevation.

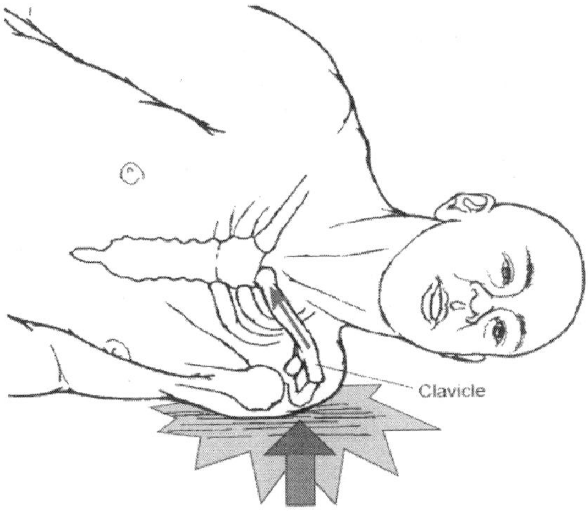

Figure 2.12. Forces that tend to medially shift the clavicle. A fall on the lateral side of the shoulder creates a force that tends to press the clavicle medially.

Source: https://www.medicalnewstoday.com/articles/321264.

Despite an intrinsically unsteady joint surface, such supporting structures collectively restrict lateral, medial, anterior, posterior, and superior clavicle displacements on the sternum, according to an analysis of the sternoclavicular joint's supporting structures. The interclavicular ligament and the costal cartilage themselves limit the clavicle's inferior mobility. As a result, the sternoclavicular joint is sufficiently strengthened to be relatively stable (Nettles and Linscheid, 1968; Thomas and Friedman, 1989).

The sternoclavicular joint, also known as a saddle or ball-and-socket joint, moves along three axes: a vertical SI, and AP, and a longitudinal (ML) axis that runs the length of the clavicle (Figure 2.13). Even though these axes are slightly oblique to the body's cardinal planes (Steindler, 1977), the clavicle's movements occur extremely near to them. Elevation and depression happen essentially in the frontal plane when moving around the AP axis.

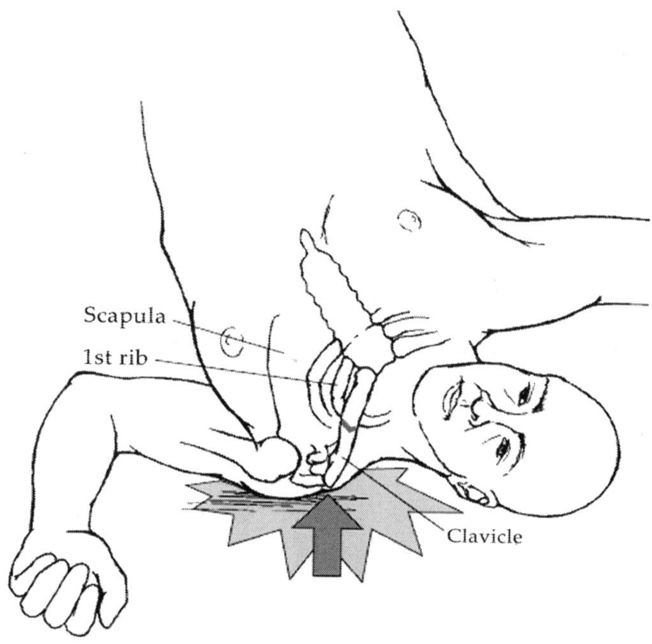

Figure 2.13. A common method of clavicle fracture.

Source: https://www.choosept.com/guide/physical-therapy-guide-collarbone-fracture-clavicle.

In the transverse plane, motions along the SI axis have been defined as retraction and protraction. Downward (anterior) and Upward (posterior) rotations have been characterized by if the anterior surface of the clavicle moves down (downward rotation) or up (upward rotation). While the sternoclavicular joint rotates, the importance of the clavicular head and the position of the joint's axis make the head of the clavicle easier to palpate during most such motions. This palpation commonly leads rookie clinicians to perplexity.

Remember that clavicle retraction induces the clavicle head to shift anteriorly on the sternum while the clavicle body moves posteriorly (Figure 2.14). Likewise, as the body advances anteriorly during protraction, the clavicular head rolls posteriorly. Similarly, when the sternoclavicular joint is in a state of elevation, the body of the acromion and the clavicle will rise, while the head of the clavicle will descend onto the sternum. When the joint is in a state of depression, the opposite will occur. The positioning of the axis within the clavicle has resulted in movements of the proximal and distal clavicular surfaces in the opposite direction from one another. These movements are compatible with rotations of the sternoclavicular joint. Although the exact location of the axes along which movement occurs in the sternoclavicular joint is unknown, it is believed that they are located laterally to the head of the clavicle (Bearn, 1967; Pronk, Van der Helm, and Rozendaal, 1993). This explains why the lateral and medial ends of the clavicle appear to move in different directions. The two ends of the clavicle move in opposite directions throughout complete rotation around the pivot point, much as the two ends of a seesaw move in opposite ways throughout full rotation around the pivot.

Research has been done to investigate the possible range of motion of the sternoclavicular joint. According to the findings of the research, the overall excursions of depression and elevation are between 50° and 60°, with depression accounting for less than 10% of the total (Moseley, 1968; Steindler, 1977). Both the elevation and the depression are restricted by the upper section of the capsule as well as the interclavicular ligament (Bearn, 1967; Stanley, Trowbridge, and Norris, 1988). The costoclavicular ligament is the one that restricts the elevation. It has been hypothesized that the clavicle's contact with the first rib can limit the depression of the sternoclavicular joint (Pronk et al., 1993). There is significant evidence for contact between the clavicle and the first costal cartilage in some patients, as indicated by the presence of facets between the two structures in certain cadaver specimens (Bearn, 1967; Pronk et al., 1993).

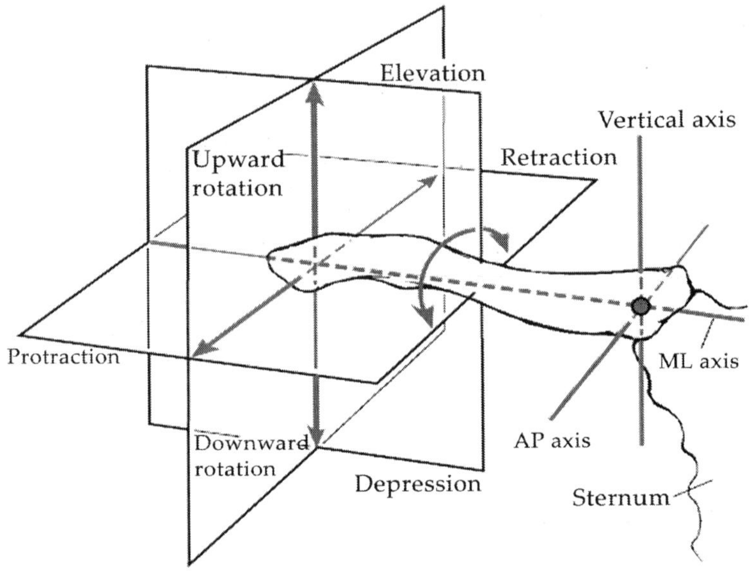

Figure 2.14. The sternoclavicular joint's axes of movement.

Source: https://ouhsc.edu/bserdac/dthompso/web/namics/scjoint.htm.

The overall excursion of retraction and protraction appears to be more comparable, spanning from 30° to 60° (Pronk et al., 1993; Stanley et al., 1988). Protraction is restricted by the posterior sternoclavicular ligament, which prevents the clavicular head from moving backwards, and the costoclavicular ligament, which prevents the clavicle's body from moving forward. The anterior sternoclavicular ligament and the costoclavicular ligament both hinder retractions. Both movements are limited by the interclavicular ligament (Bearn, 1967; Fukuda et al., 1986). Downward and upward rotations seem to be more restricted than other motions, with upward rotation ROM estimations ranging from 25° to 55° (Bearn, 1967; Boone and Azen, 1979; Fukuda et al., 1986; Inman et al., 1944; Pronk et al., 1993). While no research has been done on downward rotation ROM, it looks to be significantly smaller than upward rotation, perhaps less than 10°.

It is generally established that motion at the sternoclavicular joint is intricately tied to movements at the other joints of the shoulder complex, regardless of the particular amount of excursion accessible. After every joint is shown, the relationship between such motions is addressed.

2.3.2. Acromioclavicular Joint

In general, the acromioclavicular joint is considered to be a gliding joint having flat articular surfaces, however, the surfaces are also characterized as reciprocally convex and concave (Steindler, 1977; Williams et al., 1995) (Figure 2.15). Instead of hyaline cartilage, fibrocartilage covers both articular surfaces. A capsule strengthened inferiorly and superiorly by acromioclavicular ligaments provides support for the joint (Figure 2.16). Whereas the capsule is commonly considered as being weak, the acromioclavicular ligaments can offer the joint's principal support in cases of modest displacements and mild stresses (Fukuda et al., 1986; Lee, Debski, Chen, Woo, and Fu, 1997).

Moreover, the acromioclavicular ligaments seem to limit the posterior glide of the acromioclavicular joint independent of the extent of displacement or load (Fukuda et al., 1986). Additionally, the inferior acromioclavicular ligament can offer significant resistance to severe anterior movement of the clavicle on the scapula (Lee et al., 1997). Additionally, the joint has an intraarticular meniscus which is often smaller than a complete disc and offers no extra support.

The extracapsular coracoclavicular ligament, which extends from the base of the coracoid procedure to the inferior surface of the clavicle, provides the acromioclavicular ligament with additional support. The acromioclavicular joint relies on this ligament for stability, especially against major medial and excursion displacements (Fukuda et al., 1986; Lee et al., 1997). Many consider it to be the shoulder complex's principal suspensory ligament. It is much stiffer than the coracoacromial, acromioclavicular, and superior glenohumeral ligaments, according to mechanical testing (Costic et al., 2003).

Strangely, a ligament that does not even span the joint directly may play such a vital role in maintaining stability. Recognizing the ligament's exact orientation can assist describe its involvement in joint stabilization. The conoid ligament, which runs vertically from the coracoid procedure to the conoid tubercle on the clavicle, and the trapezoid ligament, which runs laterally and vertically to the trapezoid line, make up the ligament. The conoid ligament, which is vertically oriented, is said to restrict excessive superior glides at the acromioclavicular joint. Smaller superior displacements are supposedly limited by the acromioclavicular ligaments (Fukuda et al., 1986; Lee et al., 1997).

The trapezoid ligament defends against shearing stresses that might push the acromion medially and inferiorly beneath the clavicle. These forces may be generated by a shoulder fall or shoulder blow. Because of the form of its articular surfaces, the acromioclavicular joint is especially vulnerable to these displacements. As noted previously, the clavicle's articular facet faces inferiorly and laterally, whereas the acromion's faces superiorly and medially. Such surfaces provide the acromioclavicular joint with a beveled look, allowing for the medial displacement of the acromion behind the clavicle. Since the coracoid procedure is part of the certain scapula as the acromion, medial displacement of the acromion causes concomitant displacement of the coracoid procedure. Examining the trapezoid ligament reveals that it has been positioned to obstruct the medial translation of the coracoid procedure, so assisting in maintaining the clavicle's relationship with the scapula and avoiding dislocation (Figure 2.17) (Pronk et al., 1993). Dislocation of the acromioclavicular joint is frequently followed by rupture of the coracoclavicular ligament and fracture of the coracoid procedure.

The coracoacromial ligament is another uncommon ligament found in the acromioclavicular joint. It is uncommon since it does not cross a joint. Instead, it forms a canopy over the glenohumeral joint by connecting two scapula landmarks (Figure 2.15). This ligament offers protection to the bursa and supraspinatus tendon beneath it.

Additionally, it restricts the superior gliding of the humerus in a highly unstable glenohumeral joint (Lucas, 1973). In certain shoulders having rotator cuff tears, the coracoacromial ligament is thicker, which may play a role in the impingement of the structures that lie underneath it. The debate remains if the thickening is an outcome of contact with the unstable humerus caused by a torn rotator cuff or whether the thickening itself is a risk factor for rotator cuff tears (Soslowsky, An, Johnston, and Carpenter, 1994). The link between the anatomy of the coracoacromial ligament and the integrity of the rotator cuff muscles requires more investigation.

Certain studies provide objective assessments of the acromioclavicular joint's range of motion. During the movement of the shoulder, Sahara et al. indicate that there is a total translation of approximately 4 millimeters in the posterior and anterior directions and approximately 2 millimeters in the superior and inferior directions (Sahara, Sugamoto, Murai, Tanaka, and Yoshikawa, 2006).

Although gliding joints only permit translational motion, several publications (Culham and Peat, 1993; Pronk et al., 1993; Williams et al., 1995) report rotational motion about specified axes of motion at the acromioclavicular joint. The basic axes are vertical, AP, and medial-lateral (ML) (Figure 2.15). The vertical axis permits scapular mobility that brings the scapula nearer or further away from the clavicle in the transverse plane. In the frontal plane, movement about the AP axis enlarges or reduces the angle created by the clavicle and spine of the scapula. Movement along the ML axis tilts the superior border of the scapula toward or away from the clavicle. Direct observations of angular excursions about particular axes range from less than 10° to 20° (Inman et al., 1944; Pronk et al., 1993). Sahara et al. (2006) report a maximum of 35° of rotation including maximum shoulder abduction utilizing a screw axis (a single axis which represents the total translation and rotation). According to this research, the acromioclavicular joint permits considerable motion between the clavicle and scapula.

When the shoulder complex as a whole is considered, the acromioclavicular joint is the one that is responsible for managing the articulation of the clavicle with the scapula. This occurs even though these two bones move in distinct patterns. It is irrelevant to the clinician if this leads to systematic rotating motions or in a gliding realignment of the bones, as either in scenario the motions may not be easily detected. Even though the scapula and clavicle move together, their contributions to shoulder motion necessitate that they move relatively independently of one another (Figure 2.15).

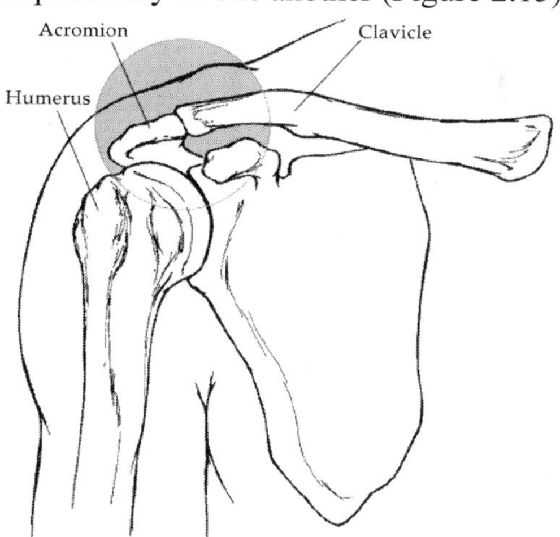

Figure 2.15. With one another, the articular surfaces of the acromioclavicular joint are comparatively beveled and flattened.

Source: https://musculoskeletalkey.com/injury-to-the-acromioclavicular-and-sternoclavicular-joints/.

2.3.3. Scapulothoracic Joint

As previously established, the scapulothoracic joint is an unconventional joint that lacks all of the standard properties of a joint save one: mobility. The fundamental function of this joint is to increase the range and variety of motions between the arm and the trunk by amplifying the motion of the glenohumeral joint. The scapulothoracic joint, together with its surrounding muscles, is also regarded as an essential shock absorber that protects the shoulder, especially during falls on an outstretched arm (Kelley and Clark, 1995). Two translations and three rotations are the primary movements of the scapulothoracic joint.

Those movements are:

- Adduction and abduction;
- Depression and elevation;
- Scapular tilt;
- External and internal rotations;
- Upward (lateral) and downward (medial) rotations.

The displacement of the whole scapula superiorly to the thorax is known as elevation. The opposite of happiness is depression. Abduction is described as moving farther from the vertebrae by the whole medial edge of the scapula, whereas adduction is expressed as a movement toward the vertebrae. The scapulothoracic joint's adduction and abduction are also known as retraction and protraction. Certain people use the term protraction to describe the coupling of abduction and upward rotation of the scapula. Others use the word protraction to describe primarily the transverse movements of the sternoclavicular joint.

Downward (medial) scapula rotation is described as a rotation along an AP axis that causes the glenoid fossa to tilt downward as the inferior angle travels toward the vertebrae. The reverse is upward (lateral) rotation. The axis of upward and downward rotation seems to be somewhat inferior to the scapular spine, almost equidistant between the vertebral and axillary boundaries (Van der Helm, 1994). The exact placement of the axis is likely to change with shoulder ROM.

The scapula rotates internally and externally along a vertical axis. Internal rotation brings the axillary border of the scapula closer to the front, whereas external rotation brings it closer to the back. This motion can be aided by the thorax's shape. The scapula rotates internally as it moves laterally on the thorax in scapular abduction. The scapula, on either hand, tends to spin externally when it adducts.

The scapula tilts anteriorly and posteriorly around the ML axis. The superior section of the scapula travels anteriorly, whereas the inferior angle of the scapula moves posteriorly.

The motion is reversed by posterior tilt. Similarly, the thorax's shape may help these actions. When the scapula rises, it tilts anteriorly, and when it depresses, it tilts posteriorly.

2.4. MUSCLE ACTIVITY IN THE SHOULDER COMPLEX: PATHOMECHANICS AND MECHANICS

The axioclavicular and axioscapular muscles all have an attachment to the axioskeleton and the shoulder girdle, i.e., the clavicle and scapula (Figure 2.16). The major function of such muscles is to move the scapulothoracic and sternoclavicular joints, leading to motion at the acromioclavicular joint. It is vital to remember that the scapula's only bone connection is at the tiny acromioclavicular joint to properly comprehend the significance of such muscles. As a result, the scapula floats loosely on the thorax, mostly supported by muscles. The axioscapular set of muscles typically works in tandem to keep the scapula stable as it rotates across the thorax. Trapezius, levator scapulae, serratus anterior, rhomboid minor and major, pectoralis minor, sternocleidomastoid, and subclavius are the muscles that make up the axioclavicular and axioscapular group. Each is addressed individually below. In addition, the team-functioning muscles are described (Figure 2.16).

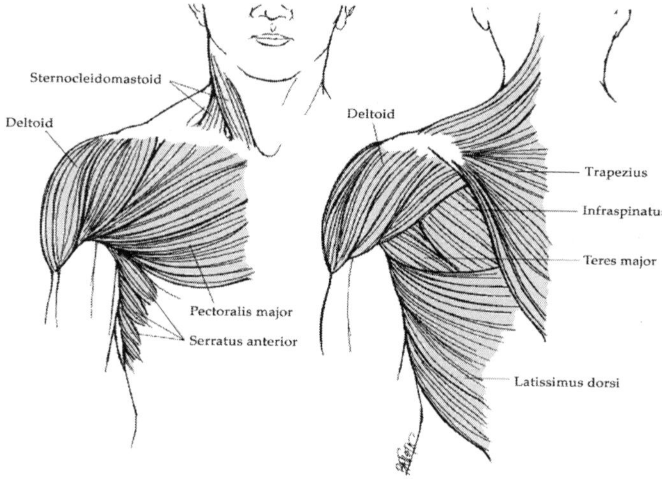

Figure 2.16. The shoulder complex's superficial muscles.

Source: https://www.physio-pedia.com/Shoulder.

2.4.1. Trapezius

There are three bellies in the trapezius: lower, middle, and upper. Each of them has a distinct role and contributes considerably to the trapezius overall function. The scapula and clavicle attachments show that the muscle operates at both the scapulothoracic and sternoclavicular joints. The actions and repercussions of every muscle belly's tightness and weakening are detailed below. The function of the entire muscle is then discussed.

The upper trapezius has substantially fewer muscular bundles than the other two parts of the trapezius, according to careful cadaver dissection of individual fascicles (Hara, Ito, and Iwasaki, 1996). According to this research, the upper trapezius fibers only connect to the clavicle, with no direct contact with the scapula.

During the active elevation of the shoulder girdle as in a shoulder shrug (Figure 2.17), electromyography (EMG) demonstrates significant activity in the upper trapezius (Boone and Azen, 1979; Chen et al., 1999; Cohen and Williams, 1998). Gradual fall of the scapula from high to upright posture is likewise followed by contraction of the upper trapezius, probably as an eccentric regulation of the motion (Chen et al., 1999; Cohen and Williams, 1998). In the same way, scapular adduction in the erect standing position activates the upper trapezius. The upper trapezius actively lifts and adducts the shoulder girdle, according to EMG findings. Even though no research has directly examined the upper trapezius muscle's involvement in scapulothoracic joint upward rotation, its connection to the lateral clavicle is compatible with that function, as the proper scapulothoracic joint upward rotation should be accompanied by clavicle elevation. EMG findings demonstrating upper trapezius activation throughout shoulder abduction indirectly suggest its importance in shoulder abduction (Figure 2.17) (Crubbs, 1993; Kebaetse et al., 1999).

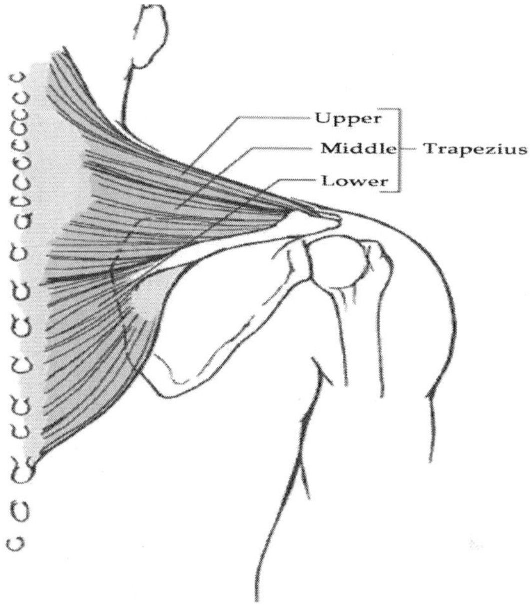

Figure 2.17. Trapezius muscle. The trapezius is split into three sections: lower, middle, and upper.

Source: https://biologydictionary.net/trapezius-muscle/.

The significance of the upper trapezius in keeping an upright posture is less clear. Although certain studies exhibit electrical activity in the upper trapezius throughout quiet upright standing, other EMG studies of the upper trapezius show little or no activity in a quiet upright posture, even though the weight of the upper extremity tries to depress the scapula and clavicle in the erect posture (Bearn, 1967; Gerhardt and Rippstein, 1990). Even the addition of hand weights, as per Basmajian, has no evident effect on the upper trapezius muscle (Bearn, 1967). Additional research is required to identify whether these discrepancies in the scientific literature reflect methodological and demographic variances across the different studies or the variety of responses observed in a normal population. Likely, the resting tension of a normal upper trapezius contributes actively to the upward stability of the shoulder complex via its connection to the clavicle (Bigliani et al., 1996; Gerhardt and Rippstein, 1990). Without a direct connection to the scapula, it is well accepted that the upper trapezius supports the shoulder girdle in an upright position (Bigliani et al., 1996; Gerhardt and Rippstein, 1990; Hara et al., 1996).

2.4.1.1. Implications of Upper Trapezial Muscle Weakness

Although isolated upper trapezius weakness is uncommon, it is a major contributor to reduced shoulder girdle elevation strength. In the existence of trapezius weakness, standing posture has been distinguished by depression, abduction, and forward tilting of the scapula, according to Steindler (Novotny, Beynnon, and Nichols, 2000). Bearn comments that even though the clavicle has been depressed utilizing trapezius paralysis, the depression is not as severe as anticipated, nor is it the only depression accessible (Bigliani et al., 1996). Even if research described in the preceding section refutes active upper trapezius contributions to an erect posture, the postural anomalies generally linked with upper trapezius weakening can originate from a lack of adequate resting tone in a weaker upper trapezius. Furthermore, they can also be the consequence of overall trapezius muscular weakening. Additional research is required to establish a definite association between postural alignment and upper trapezius strength.

2.4.1.2. Effects of Tightness of the Upper Trapezius

Upper trapezius tightness is linked to raised shoulders, asymmetrical head postures, and limited neck and head ranges of motion (ROM). But, because alternative scapular elevators are present, determining pure upper trapezius tightness is challenging. The scapular elevation is likely to be followed by an upward rotation of the scapula if the upper trapezius is tight alone. To distinguish between the scapular elevators, a detailed examination of scapular posture is required.

Due to its horizontally oriented fibers, the middle trapezius is considered a pure scapular adductor. The middle trapezius has the biggest cross-sectional area of the three trapezius muscle sections in cadaver specimens (Hara et al., 1996). As a result, the middle trapezius offers significant strength in scapular adduction and helps to stabilize the scapula. The upper trapeziu's top fibers may aid the small upper trapezius in scapular elevation, according to EMG research (Crubbs, 1993).

2.4.1.3. Weakness of the Middle Trapezius

Significantly diminished scapular adduction strength is the consequence of middle trapezius weakness. While the isolated weakening of the middle trapezius is uncommon, certain writers believe that it may result from prolonged muscular stretching in a posture typified by scapular abduction (Hawkins, Schutte, Janda, and Huckell, 1996). Yet, tries to connect scapular position with middle trapezius strength (Couteau et al., 2000) have failed. Reduction in strength in the middle trapezius also impedes the contraction of the scapulohumeral muscles. To apply their force at the glenohumeral joint, the lateral rotators of the shoulder, comprising the posterior deltoid and infraspinatus muscles, need a stable scapula.

Weakness in scapular adduction allows such muscles to draw the scapula toward the humerus, rather than the humerus toward the scapula.

2.4.1.4. Tightness of the Middle Trapezius

Since the weight of the whole upper limb forces the scapula into abduction, the rigidity of the middle trapezius alone is uncommon. A thorough examination of the line of pull of the lower trapezius reveals the muscle's potential contributions to each of these motions. This analysis also reveals that the muscle is optimally suited for upward rotation and depression. The line of tension of the lower trapezius is optimal for scapula depression.

Nevertheless, in an erect position, the weight of the upper extremities already drags the scapula downward. Extra depression from lower trapezius activation is unnecessary. In contrast, manual resistance against scapular depression can stimulate the electrical activity of the inferior trapezius when the individual is prone (Boone and Azen, 1979). The significance of the scapular depression force supplied by the lower trapezius is most evident when the upper trapezius is simultaneously contracted. This integrated action is described when discussing the trapezius as a whole.

The lower trapezius's EMG activity during isometric shoulder adduction from the abducted posture can also support its involvement as a scapular adductor (Crubbs, 1993). It is necessary to recollect the placement of the axis for downward and upward scapular rotation to comprehend the role of the inferior trapezius muscle in upward scapular rotation. Even though the exact placement of the axis is unknown, it is obvious that it is located laterally to the lower trapeziu's scapular connection on the root of the scapula's spine. As a result, the scapula rotates and rises as the lower trapezius pushes the medial portion of the scapular spine inferiorly (Figure 2.17). The lower trapeziu's activity throughout shoulder elevation, such as the upper trapeziu's,' supports its contributions as an upward rotator of the scapula.

2.4.1.5. Weakness of the Lower Trapezius

A prolonged stretch caused by an elevated and downwardly rotated scapula is considered the cause of isolated lower trapezius weakness (Hawkins et al., 1996). Furthermore, no studies have been found to support this theory. If the lower trapezius is weak, it may be hard to stabilize the scapula throughout the contraction of the other upward rotators.

2.4.1.6. Tightness of the Lower Trapezius

Theoretically, stiffness of the lower trapezius leads to reduced elevation and downward rotation ROM of the scapulothoracic joint, as well as a possible depressed and posteriorly inclined shoulder girdle in silent standing. There have been no known examples of unilateral lower trapezius stiffness; although, healthy persons frequently report differences in shoulder height. Such difference is related to hand dominance (Hawkins et al., 1996; Nettles and Linscheid, 1968). There are various plausible explanations for the lack of lower trapezius tightness despite the apparent scapular depression in certain individuals. The depression can be followed by simultaneous scapular downward rotation and/or abduction, which may counteract the shortening impact of the depression or perhaps surpass it. Despite the extended scapular depression, there can be no adaptive modification in the lower trapezius since it is sufficiently disrupted by scapular elevation throughout upper extremity usage. Additionally, because there has been no generally acknowledged norm of scapular posture in healthy persons, what looks to be scapular depression may instead be contralateral elevation. Consequently, the limitations caused by stiffness in the lower trapezius are unknown and may not exist.

2.4.1.7. Actions of the Entire Trapezius

The vector sum of the forces from the lower, middle, and upper trapezius muscles may be regarded as the activities of the entire trapezius. The trapezius adducts and rotates the scapula upwards as a whole. The lower and upper component's elevation and depression motions, correspondingly, balance one another out. In reality, the scapula's stability is dependent on this balancing of two opposing forces.

The *anatomical force pair* is formed by such two muscles pulling in different directions and generating rotation of the scapula. The lower and upper trapezius muscles work together to allow the scapula to rotate upward without moving inferiorly and superiorly on the thorax. An imbalance between such two muscles, caused by stiffness or weakness in one of them, may make it very hard to stabilize the scapula during scapulothoracic joint upward rotation (such as during the abduction or flexion of the shoulder).

The trapezius as a whole contributes significantly to the scapular upward rotation required for proper arm–trunk abduction or flexion. Shoulder abduction seems to be more important than shoulder flexion (Gerhardt and Rippstein, 1990; Kebaetse et al., 1999). The fact that the muscle is largely in the frontal plane explains its increased significance in shoulder abduction. The complete trapezius adducts and rotates the scapula upwards on its own. However, without considerable scapular adduction, normal arm-trunk elevation occurs. As a result, another muscle is required to balance the adduction element of the trapezius. The serratus anterior provides this equilibrium.

2.4.2. The Mechanical Demands on Shoulder Complex Structures

Researchers have worked hard to develop models that better reflect the anatomy and behavior of the entire shoulder complex (Bagg and Forrest, 1988; Bigliani et al., 1996; Borsa, Sauers, Herling, and Manzour, 2001; Costic et al., 2003; Couteau et al., 2000; Deutsch et al., 1996). As previously noted, the traditional method for analyzing forces in a joint is to employ basic assumptions to restrict the number of unknowns to the number of equations accessible to explain the phenomenon. Moreover, the fact is that there are many more unfamiliar than equations at any joint in the human body. This is referred to as *redundancy*, and the system with these unknowns is considered to be *indeterminate*, with an endless number of equation solutions. Furthermore, advanced mathematical methods exist that enable researchers to find the "best" or "optimal" option based on a set of predefined optimization criteria. Various models are constructed utilizing this method to identify the forces in the glenohumeral joint's muscles and ligaments, and the other joints of the shoulder complex. The data from certain such models are presented briefly below. Such figures are only estimations of the actual loads that the shoulder constructions must withstand. They may, however, give the clinician some insight into the demands and consequences of activities and exercise on separate shoulder complex structures.

Van derHelm (Regulation, 2002) estimates peak ML joint reaction forces at the glenohumeral joint of roughly 100 newtons (22.5 lb) and 300 newtons (67 pounds) for flexion and abduction, correspondingly, in a far more complicated model of the shoulder. (One kilogram is 9.81 newtons (N), while one pound is 4.45 N.)

The reaction forces at the AP and longitudinal joints are marginally reduced. Although prior research evaluated joint reaction forces, now this displays the individual 3D components instead of the total response force. As a result, the magnitudes may not be directly compared.

The peaks rise at around 90°, as in earlier research, when the moment owing to upper extremity weight is greatest. Peak joint response forces in the acromioclavicular and sternoclavicular joints were reported to be around 50 and 120 newtons (11 and 27 pounds) in the lateral-medial direction in the same study. The importance of the scapulothoracic joint in the stability and mobility of the entire shoulder complex is demonstrated in this study. It is also one of the only studies to give any indication of the loads maintained at the shoulder's other joints. On abduction with one-kilogram weight in the hand, a comparable, albeit less comprehensive model assumes muscle loads up to 150 (34 pounds) in the deltoid and above 100 (22.5 lb) in the supraspinatus (Borsa et al., 2001). In the middle of the range of motion, this study reveals peak joint response forces of around 80% of body weight.

Until now, mathematical evaluations based on normal Newtonian mechanics have been utilized to predict the loads on the soft tissues and joints of the shoulder complex. An alternative method employs anatomically-based models that simulate muscle behavior. By developing accurate physical models of shoulder muscles or analyzing the relative activity of shoulder muscles, such studies shed light on the relative difficulty of activities and the roles of various muscles in such activities (Ludewig and Cook, 2000; McQuade, Dawson, and Smidt, 1998; Murray, Gore, Gardner, and Mollinger, 1985).

According to one such study, the average peak force needed by the deltoid muscle to abduct the shoulder in the plane of the scapula is roughly 250 newton (34.5 pounds), or roughly 56 pounds (Freedman and Munro,

1966). Also reported are the impacts of the absence of the supraspinatus, the entire rotator cuff, and the deltoid muscles. When the supraspinatus, the other rotator cuff muscles, or the deltoid is selectively severed, the total active shoulder elevation reduces by 6, 16, and 25%, respectively. According to a comparable model employing a physical model of the shoulder muscles, the force needed by the deltoid to raise the upper extremity in the absence of the rotator cuff muscles rises by 17% (Campbell, 1992; Cooper, 1992). Another study employs EMG and intramuscular pressure to assess the relative activity of the supraspinatus and confirms what mathematical calculations have shown, namely, that the activity of the supraspinatus rises when the shoulder is abducted between 0° and 90° (Boone and Azen, 1979). Such studies aid in identifying the relative contributions and relevance of specific shoulder structures to the shoulder's overall function. They consistently show that the loss of function in a single muscle leads to either a reduction in motion or a rise in the load carried by the remaining muscles. Such findings give a theoretical foundation for explaining clinical observations like a patient's complaints of shoulder weakness and fatigue when a rotator cuff injury is present.

Two further studies demonstrate how research into muscle loads may yield information that can help clinicians better comprehend the mechanical basis for a patient's complaints. To test the fatigability of certain such muscles, the first study employs a mathematical model of the shoulder and EMG recordings of the shoulder muscles (Couteau et al., 2000). The infraspinatus, deltoid, and supraspinatus muscles, according to these authors, are the 1st to exhibit indications of exhaustion after continuous isometric shoulder contractions at 90° of flexion against an a^4-kg weight. In this study, the trapezius was found to be more fatigue resilient. The 2nd study looks at the degree of EMG activity in shoulder muscles in various elbow and shoulder positions, both without or with a lower-resistance manual job (Dayanidhi et al., 2005). When manual tasks are present, EMG activity increases in practically all locations and muscles. While neither study directly measures the forces created in the shoulder muscles, they do imply that even slight increases in the weights carried by the upper extremities can considerably raise the loads endured by muscles.

2.5. THE RELATIONSHIP BETWEEN JOINT AND MUSCLE FORCE ANALYSIS AND CLINICAL PRACTICE

The findings of much research into the forces experienced by the muscles and joints of the shoulder complex are also covered here. Simplifying hypotheses or simple physical models of complex anatomical features are used in the analysis. Consequently, such data are the most accurate representation of the shoulder's actual loads. When the absolute values of forces recorded in such trials are compared to the maximal loads for muscle, cartilage, and bone the doctor may evaluate the activities or exercise's potential for harm (Hislop and Montgomery, 1995; Jv and De Luca, 1985).

Likewise, similar knowledge is necessary for the development of appropriate joint replacement devices. In a wider sense, this research provides a theoretical foundation for clinicians to assess any patient's problems. Even a crude model of the forces involved in the process permits the doctor to ask, "How much muscular effort is needed to lift this 20-pound baby?" and, maybe more crucially, is there a different way to raise the baby that requires less muscle force? Likewise, the clinician may inquire, "How much load is being applied to this inflammatory joint throughout this strengthening exercise?" Is there a way to modify the workout to minimize the impact on the joint? While few doctors have the chance to respond to these questions statistically, knowing the basic method of the study allows the therapist to produce hypothetical solutions. Clinical findings may then confirm or disprove these predictions.

REFERENCES

1. Bagg, S. D., & Forrest, W. J., (1988). A biomechanical analysis of scapular rotation during arm abduction in the scapular plane. *Am. J. Phys. Med. Rehabil., 67*(6), 238–245.
2. Bearn, J., (1967). Direct observations on the function of the capsule of the sternoclavicular joint in clavicular support. *Journal of Anatomy, 101*(1), 159.
3. Ben, K. W., (1998). The role of the scapula in athletic shoulder function. *The American Journal of Sports Medicine, 26*(2), 325–337.
4. Bigliani, L. U., Kelkar, R., Flatow, E. L., Pollock, R. G., & Mow, V. C., (1996). Glenohumeral stability: Biomechanical properties of passive and active stabilizers. *Clinical Orthopedics and Related Research (1976–2007), 330*, 13–30.
5. Bigliani, L., (1986). The morphology of the acromion and its relationship to rotator cuff tears. *Orthop Trans., 10*, 228.
6. Boone, D. C., & Azen, S. P., (1979). Normal range of motion of joints in male subjects. *JBJS, 61*(5), 756–759.
7. Borsa, P. A., Sauers, E. L., Herling, D. E., & Manzour, W. F., (2001). *In vivo* quantification of capsular end point in the nonimpaired glenohumeral joint using an instrumented measurement system. *Journal of Orthopedic & Sports Physical Therapy, 31*(8), 419–431.
8. Branch, T. P., Lawton, R. L., lobst, C. A., & Hutton, W. C., (1995). The role of glenohumeral capsular ligaments in internal and external rotation of the humerus. *The American Journal of Sports Medicine, 23*(5), 632–637.
9. Campbell, W. C., (1992). *Campbell's Operative Orthopedics* (Vol. 4): Mosby-Year Book.
10. Chen, S. K., Simonian, P. T., Wickiewicz, T. L., Otis, J. C., & Warren, R. F., (1999). Radiographic evaluation of glenohumeral kinematics: A muscle fatigue model. *Journal of Shoulder and Elbow Surgery, 8*(1), 49–52.
11. Cohen, R. B., & Williams, Jr. G. R., (1998). Impingement syndrome and rotator cuff disease as repetitive motion disorders. *Clinical Orthopedics and Related Research,* (351), 95–101.
12. Cooper, R. R., (1992). Campbell's operative orthopedics. *JAMA, 268*(13), 1779–1780.

13. Costic, R. S., Vangura, Jr. A., Fenwick, J. A., Rodosky, M. W., & Debski, R. E., (2003). Viscoelastic behavior and structural properties of the coracoclavicular ligaments. *Scandinavian Journal of Medicine & Science in Sports, 13*(5), 305–310.
14. Couteau, B., Mansat, P., Darmana, R., Mansat, M., & Egan, J., (2000). Morphological and mechanical analysis of the glenoid by 3D geometric reconstruction using computed tomography. *Clinical Biomechanics, 15*, S8–S12.
15. Crubbs, N., (1993). Frozen shoulder syndrome: A review of literature. *Journal of Orthopedic & Sports Physical Therapy, 18*(3), 479–487.
16. Culham, E., & Peat, M., (1993). Functional anatomy of the shoulder complex. *Journal of Orthopedic & Sports Physical Therapy, 18*(1), 342–350.
17. Curl, L. A., & Warren, R. F., (1996). Glenohumeral joint stability: Selective cutting studies on the static capsular restraints. *Clinical Orthopedics and Related Research (1976–2007), 330*, 54–65.
18. Dayanidhi, S., Orlin, M., Kozin, S., Duff, S., & Karduna, A., (2005). Scapular kinematics during humeral elevation in adults and children. *Clinical Biomechanics, 20*(6), 600–606.
19. Deutsch, A., Altchek, D. W., Schwartz, E., Otis, J. C., & Warren, R. F., (1996). Radiologic measurement of superior displacement of the humeral head in the impingement syndrome. *Journal of Shoulder and Elbow Surgery, 5*(3), 186–193.
20. Ebaugh, D. D., McClure, P. W., & Karduna, A. R., (2005). Three-dimensional scapulothoracic motion during active and passive arm elevation. *Clinical Biomechanics, 20*(7), 700–709.
21. Field, L. D., Warren, R. F., O'Brien, S. J., Altchek, D. W., & Wickiewicz, T. L., (1995). Isolated closure of rotator interval defects for shoulder instability. *The American Journal of Sports Medicine, 23*(5), 557–563.
22. Flatow, E. L., Duralde, X. A., Nicholson, G. P., Pollock, R. G., & Bigliani, L. U., (1995). Arthroscopic resection of the distal clavicle with a superior approach. *Journal of Shoulder and Elbow Surgery, 4*(1), 41–50.
23. Freedman, L., & Munro, R. R., (1966). Abduction of the arm in the scapular plane: Scapular and glenohumeral movements: A roentgenographic study. *JBJS, 48*(8), 1503–1510.

24. Fukuda, K., Craig, E., An, K., Cofield, R., & Chao, E., (1986). Biomechanical study of the ligamentous system of the acromioclavicular joint. *The Journal of Bone and Joint Surgery; American, 68*(3), 434–440.
25. Gerhardt, J. J., & Rippstein, J., (1990). *Measuring and Recording of Joint Motion: Instrumentation and Techniques*. Hans Huber Pub.
26. GJE, R., (1981). *Cunningham's Textbook of Anatomy*. Oxford: Oxford University Press.
27. Graichen, H., Stammberger, T., Bonel, H., Englmeier, K. H., Reiser, M., & Eckstein, F., (2000). Glenohumeral translation during active and passive elevation of the shoulder—a 3D open-MRI study. *Journal of Biomechanics, 33*(5), 609–613.
28. Graichen, H., Stammberger, T., Bonel, H., Haubner, M., Englmeier, K. H., Reiser, M., & Eckstein, F., (2000). Magnetic resonance based motion analysis of the shoulder during elevation. *Clinical Orthopedics and Related Research (1976–2007), 370*, 154–163.
29. Hara, H., Ito, N., & Iwasaki, K., (1996). Strength of the glenoid labrum and adjacent shoulder capsule. *Journal of Shoulder and Elbow Surgery, 5*(4), 263–268.
30. Harrington, Jr. M., Keller, T., Seiler, III. J., Weikert, D., Moeljanto, E., & Schwartz, H., (1993). Geometric properties and the predicted mechanical behavior of adult human clavicles. *Journal of Biomechanics, 26*(4, 5), 417–426.
31. Harryman 2nd, D., Sidles, J., Clark, J. M., McQuade, K. J., Gibb, T. D., & Matsen, 3rd, F., (1990). Translation of the humeral head on the glenoid with passive glenohumeral motion. *JBJS, 72*(9), 1334–1343.
32. Harryman, D., Sidles, J., Harris, S., & Matsen, F., (1992). The role of the rotator interval capsule in passive motion and stability of. *J. Bone Joint Surg. Am., 74*, 53–66.
33. Hawkins, R. J., Schutte, J. P., Janda, D. H., & Huckell, G. H., (1996). Translation of the glenoihumeral joint with the patient under anesthesia. *Journal of Shoulder and Elbow Surgery, 5*(4), 286–292.
34. Hislop, J., & Montgomery, J., (1995). *Daniels and Worthinghams Muscle Testing: Techniques of Manual Examination*. SB Saunders company. Philadelphia, Pennsylvania [Google Scholar].

35. Hjelm, R., Draper, C., & Spencer, S., (1996). Anterior-inferior capsular length insufficiency in the painful shoulder. *Journal of Orthopedic & Sports Physical Therapy, 23*(3), 216–222.
36. Howell, S. M., & Galinat, B. J., (1989). The glenoid–labral socket: A constrained articular surface. *Clinical Orthopedics and Related Research®, 243*, 122–125.
37. Iannotti, J. P., Gabriel, J. P., Schneck, S., Evans, B., & Misra, S., (1992). The normal glenohumeral relationships. An anatomical study of one hundred and forty shoulders. *The Journal of Bone and Joint Surgery; American, 74*(4), 491–500.
38. Inman, V. T., Saunders, J. D. M., & Abbott, L. C., (1944). Observations on the function of the shoulder joint. *JBJS, 26*(1), 1–30.
39. Inokuchi, W., Olsen, B. S., Søjbjerg, J. O., & Sneppen, O., (1997). The relation between the position of the glenohumeral joint and the intraarticular pressure: An experimental study. *Journal of Shoulder and Elbow Surgery, 6*(2), 144–149.
40. Itoi, E., Berglund, L. J., Grabowski, J. J., Naggar, L., Morrey, B. F., & An, K. N., (1998). *Superior-Inferior Stability of the Shoulder: Role of the Coracohumeral Ligament and the Rotator Interval Capsule.* Paper presented at the Mayo Clinic Proceedings.
41. Jobe, C. M., & Lannotti, J. P., (1995). Limits imposed on glenohumeral motion by joint geometry. *Journal of Shoulder and Elbow Surgery, 4*(4), 281–285.
42. Jobe, C. M., (1997). Superior glenoid impingement. *Orthopedic Clinics, 28*(2), 137–143.
43. Jv, B., & De Luca, C., (1985). *Muscles Alive: Their Functions Revealed by Electromyography.*
44. Kapandji, I., (1982). *The Physiology of the Joints—Upper Limb* (Vol. 1). Edinburgh: Churchill-Livingstone.
45. Karduna, A. R., Williams, G. R., Williams, J. L., & Iannotti, J. P., (1997). Glenohumeral joint translations before and after total shoulder arthroplasty. A study in cadavera. *JBJS, 79*(8), 1166–1174.
46. Kebaetse, M., McClure, P., & Pratt, N. A., (1999). Thoracic position effect on shoulder range of motion, strength, and three-dimensional scapular kinematics. *Archives of Physical Medicine and Rehabilitation, 80*(8), 945–950.

47. Kelley, M. J., & Clark, W. A., (1995). *Orthopedic Therapy of the Shoulder*: Lippincott Williams & Wilkins.

48. Kennedy, J. C., & Cameron, H., (1954). Complete dislocation of the acromio-clavicular joint. *The Journal of Bone and Joint Surgery; British, 36*(2), 202–208.

49. Kent, B. E., (1971). Functional anatomy of the shoulder complex: A review. *Physical Therapy, 51*(8), 867–888.

50. Kuhn, J. E., Huston, L. J., Soslowsky, L. J., Shyr, Y., & Blasier, R. B., (2005). External rotation of the glenohumeral joint: Ligament restraints and muscle effects in the neutral and abducted positions. *Journal of Shoulder and Elbow Surgery, 14*(1), S39–S48.

51. Lee, K. W., Debski, R. E., Chen, C. H., Woo, S. L. Y., & Fu, F. H., (1997). Functional evaluation of the ligaments at the acromioclavicular joint during anteroposterior and superoinferior translation. *The American Journal of Sports Medicine, 25*(6), 858–862.

52. Levinsohn, E. M., & Santelli, E. D., (1991). Bicipital groove dysplasia and medial dislocation of the biceps brachii tendon. *Skeletal Radiology, 20*(6), 419–423.

53. Lewis, M. M., Ballet, F. L., Kroll, P. G., & Bloom, N., (1985). En bloc clavicular resection: Operative procedure and postoperative testing of function. Case reports. *Clinical Orthopedics and Related Research,* (193), 214–220.

54. Lucas, D. B., (1973). Biomechanics of the shoulder joint. *Archives of Surgery, 107*(3), 425–432.

55. Ludewig, P. M., & Cook, T. M., (2000). Alterations in shoulder kinematics and associated muscle activity in people with symptoms of shoulder impingement. *Physical Therapy, 80*(3), 276–291.

56. Lukasiewicz, A. C., McClure, P., Michener, L., Pratt, N., & Sennett, B., (1999). Comparison of 3-dimensional scapular position and orientation between subjects with and without shoulder impingement. *Journal of Orthopedic & Sports Physical Therapy, 29*(10), 574–586.

57. Magermans, D., Chadwick, E., Veeger, H., & Van, D. H. F., (2005). Requirements for upper extremity motions during activities of daily living. *Clinical Biomechanics, 20*(6), 591–599.

58. McClure, P. W., Michener, L. A., Sennett, B. J., & Karduna, A. R., (2001). Direct 3-dimensional measurement of scapular kinematics

during dynamic movements *in vivo. Journal of Shoulder and Elbow Surgery, 10*(3), 269–277.

59. McQuade, K. J., & Smidt, G. L., (1998). Dynamic scapulohumeral rhythm: The effects of external resistance during elevation of the arm in the scapular plane. *Journal of Orthopedic & Sports Physical Therapy, 27*(2), 125–133.
60. McQuade, K. J., Dawson, J., & Smidt, G. L., (1998). Scapulothoracic muscle fatigue associated with alterations in scapulohumeral rhythm kinematics during maximum resistive shoulder elevation. *Journal of Orthopedic & Sports Physical Therapy, 28*(2), 74–80.
61. Michener, L. A., McClure, P. W., & Karduna, A. R., (2003). Anatomical and biomechanical mechanisms of subacromial impingement syndrome. *Clinical Biomechanics, 18*(5), 369–379.
62. Moseley, H., & Övergaard, B., (1962). The anterior capsular mechanism in recurrent anterior dislocation of the shoulder. *The Journal of Bone and Joint Surgery; British, 44*(4), 913–927.
63. Moseley, H., (1968). 3 the clavicle: Its anatomy and function. *Clinical Orthopedics and Related Research®, 58*, 17–28.
64. Murray, M., Gore, D., Gardner, G., & Mollinger, L., (1985). Shoulder motion and muscle strength of normal men and women in two age groups. *Clinical Orthopedics and Related Research,* (192), 268–273.
65. Nettles, J. L., & Linscheid, R. L., (1968). Sternoclavicular dislocations. *Journal of Trauma and Acute Care Surgery, 8*(2), 158–164.
66. Neumann, C., Petersen, S., & Jahnke, A., (1991). MR imaging of the labral-capsular complex: Normal variations. *American Journal of Roentgenology (AJR), 157*(5), 1015–1021.
67. Neviaser, T. J., (1987). Adhesive capsulitis. *The Orthopedic clinics of North America, 18*(3), 439–443.
68. Nk, Poppen., & Walker, P. S., (1976). Normal and abnormal motion of the shoulder. *J. Bone Joint Surg. A, 58*, I95.
69. Norkin, C. C., & White, D. J., (2016). *Measurement of Joint Motion: A Guide to Goniometry*: FA Davis.
70. Novotny, J. E., Beynnon, B. D., & Nichols, C. E., (2000). Modeling the stability of the human glenohumeral joint during external rotation. *Journal of Biomechanics, 33*(3), 345–354.
71. O'Connell, P. W., Nuber, G. W., Mileski, R. A., & Lautenschlager, E., (1990). The contribution of the glenohumeral ligaments to anterior

stability of the shoulder joint. *The American Journal of Sports Medicine, 18*(6), 579–584.

72. Pagnani, M. J., Deng, X. H., Warren, R. F., Torzilli, P. A., & Altchek, D. W., (1995). Effect of lesions of the superior portion of the glenoid labrum on glenohumeral translation. *The Journal of Bone and Joint Surgery; American, 77*(7), 1003–1010.
73. Pratt, N. E., (1994). Anatomy and biomechanics of the shoulder. *Journal of Hand Therapy, 7*(2), 65–76.
74. Pronk, G., Van, D. H. F., & Rozendaal, L., (1993). Interaction between the joints in the shoulder mechanism: The function of the costoclavicular, conoid and trapezoid ligaments. *Proceedings of the Institution of Mechanical Engineers, Part H: Journal of Engineering in Medicine, 207*(4), 219–229.
75. Regulation, A., (2002). *Standards of Medical Fitness(pp. 1-9).*
76. Roberts, D., (1974). Structure and function of the primate scapula. *Primate Locomotion,* 171–200.
77. Saha, A., (1973). Mechanics of elevation of glenohumeral joint. *Acta Orthop Scand.*
78. Saha, A., Das, A., & Dutta, S., (1983). Mechanism of shoulder movements and a plea for the recognition of "zero position" of glenohumeral joint. *Clinical Orthopedics and Related Research®, 173*, 3–10.
79. Sahara, W., Sugamoto, K., Murai, M., Tanaka, H., & Yoshikawa, H., (2006). 3D kinematic analysis of the acromioclavicular joint during arm abduction using vertically open MRI. *Journal of Orthopedic Research, 24*(9), 1823–1831.
80. Soslowsky, L. J., An, C. H., Johnston, S. P., & Carpenter, J. E., (1994). Geometric and mechanical properties of the coracoacromial ligament and their relationship to rotator cuff disease. *Clinical Orthopedics and Related Research,* (304), 10–17.
81. Soslowsky, L. J., Malicky, D. M., & Blasier, R. B., (1997). Active and passive factors in inferior glenohumeral stabilization: A biomechanical model. *Journal of Shoulder and Elbow Surgery, 6*(4), 371–379.
82. Soslowsky, L., Flatow, E., Bigliani, L., Pawluk, R., Ateshian, G., & Mow, V., (1992). Quantitation of in situ contact areas at the glenohumeral joint: A biomechanical study. *Journal of Orthopedic Research, 10*(4), 524–534.

83. Spencer, Jr. E. E., Valdevit, A., Kambic, H., Brems, J. J., & Iannotti, J. P., (2005). The effect of humeral component anteversion on shoulder stability with glenoid component retroversion. *JBJS, 87*(4), 808–814.
84. Stanley, D., Trowbridge, E., & Norris, S., (1988). The mechanism of clavicular fracture. A clinical and biomechanical analysis. *The Journal of Bone and Joint Surgery; British, 70*(3), 461–464.
85. Steindler, A., (1977). *Kinesiology of the Human Body Under Normal and Pathological Conditions*. Spring-field, IL. Charles C Thomas.
86. Stokdijk, M., Eilers, P., Nagels, J., & Rozing, P., (2003). External rotation in the glenohumeral joint during elevation of the arm. *Clinical Biomechanics, 18*(4), 296–302.
87. Terry, G. C., Hammon, D., France, P., & Norwood, L. A., (1991). The stabilizing function of passive shoulder restraints. *The American Journal of Sports Medicine, 19*(1), 26–34.
88. Thomas, Jr. C. B., & Friedman, R. J., (1989). Ipsilateral sternoclavicular dislocation and clavicle fracture. *Journal of Orthopedic Trauma, 3*(4), 355–357.
89. Van, D. H. F. C., & Pronk, G. M., (1995). *Three-Dimensional Recording and Description of Motions of the Shoulder Mechanism.(Vol. 1, pp. 4-8)*
90. Van, D. H. F. C., (1994). A finite element musculoskeletal model of the shoulder mechanism. *Journal of Biomechanics, 27*(5), 551–569.
91. Van, D. H. F. C., Veeger, H., Pronk, G., Van, D. W. L., & Rozendal, R., (1992). Geometry parameters for musculoskeletal modelling of the shoulder system. *Journal of Biomechanics, 25*(2), 129–144.
92. Williams, P. L., Barnistar, L., Berry, M., Collin, P., DM, D., & JE, F. M., (1995). *Gray's anatomy, the Anatomical Basis of Medicine and Surgery* (38th British edn., p. 932). New–York, Churchill, Livingstone.
93. Williams, W., Baltimore, Belenkii, V., Gurnkel, V. S., & Paltsev, Y., (1967) *Elements of Control of Voluntary Movements* (p. 3946). Biozika, 12, 135141Carlson.
94. Wuelker, N., Korell, M., & Thren, K., (1998). Dynamic glenohumeral joint stability. *Journal of Shoulder and Elbow Surgery, 7*(1), 43–52.
95. Yanai, T., Fuss, F. K., & Fukunaga, T., (2006). *In vivo* measurements of subacromial impingement: Substantial compression develops in abduction with large internal rotation. *Clinical Biomechanics, 21*(7), 692–700.

CHAPTER 3

KINESIOLOGY OF ELBOW UNITS

CONTENTS

3.1. Introduction 76

3.2. Structure of the Bones of Elbow 76

3.3. Articulations and Supporting Structures of the Elbow 84

3.4. Mechanics and Pathomechanics of Muscle Activity at the Elbow 87

3.5. Comparisons Among the Elbow Flexors 91

References 97

3.1. INTRODUCTION

The anatomy and functioning of the shoulders were discussed in the preceding chapter. It is demonstrated that the objective of the shoulder, to place the forearms in space, necessitates extraordinary flexibility in the shoulder joint. The unusual cooperation of four different joints, and also the extraordinary mobility accessible at the glenohumeral joint, enable this flexibility. This mobility, though, means the loss of consistency. The glenohumeral joint includes various distinct anatomical characteristics and structures that contribute to its strength, most notably the rotator cuff muscle (Bowden, 1983; Dalton, Jakobi, Allman, and Rice, 2010; Solomonow et al., 1990).

The role of the elbow, on the other hand, is easier. The elbow's primary function is to reduce or extend the forearms, enabling the hand to shift to and fro from the body while performing tasks like going into the refrigerator and delivering a snack to the mouth. Furthermore, the elbow helps transform the hand to and fro from the body. Reduced structural diversity corresponds to these simpler needs. When accessible movement is reduced, intrinsic steadiness increases significantly. That chapter describes the functional and structural needs of the elbow joint, as well as how movement and stable difficulties at the elbow vary from those at the shoulder (Garland, Enoka, Serrano, and Robinson, 1994; Keskula, 2004).

3.2. STRUCTURE OF THE BONES OF ELBOW

The articulations between the distal humerus, proximal ulna, and proximal radius make up the elbow joint. Here are the important facts for every bone. Just the features of every bone which pertain to the elbow are covered, as in the previous section on the shoulder complex. As a result, the distal humerus, proximal radius, and ulna are discussed in this chapter (Figure 3.1) (Bunata, Brown, and Capelo, 2007; Callaway et al., 1997).

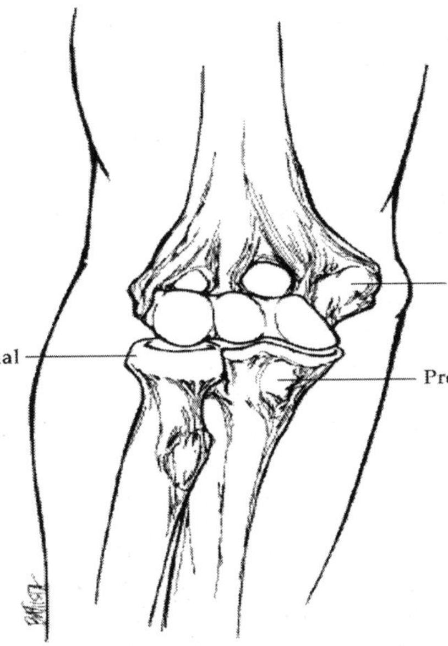

Figure 3.1. The distal humerus, proximal ulna, and proximal radius make up the elbow joint complexity.

Source: https://musculoskeletalkey.com/structure-and-function-of-the-elbow-and-forearm-complex/.

The humerus up to the deltoid tuberosity and radial groove is found in the humerus' midshaft. The humerus' shaft is about spherical in cross-sections proximally, as well as it progresses distally, it straightens anteriorly and posteriorly as well as expands medially plus laterally (Burkhart et al., 2011). The distal shaft also bends somewhat anteriorly, guiding the articular interfaces further anteriorly and favoring flexion movement. The lateral and medial supracondylar grooves are shaped by the distal plane of the humerus flattening (Anglen, 2005).

The articular surface, which includes the trochlea and capitulum, as well as the non-articulating surfaces, which include the lateral and medial epicondyles, and the olecranon fossa as well as the coronoid as well as radial fossae, make up the proximal end of the humerus (Shiba et al., 1988; Wadsworth, 1964). The medial and lateral epicondyles are conspicuous projections from the lateral and medial supracondylar ridges, respectively. The medial epicondyle is much more pronounced as compared to the lateral epicondyle, even though both are palpable.

About a third of the distal humerus is covered by it. The ulnar nerve is perforated posteriorly by a small sulcus. Because the ulnar nerve runs directly against the bone inside this channel, it may be compressed towards the humerus by a blow to the medial elbow fascial roof runs either from the medial epicondyle to the proximal end of the ulna's olecranon procedure, covering the groove. The cubital tunnel for the ulnar nerve is formed by this roof (Figure 3.2) (Polatsch, Melone, Beldner, and Incorvaia, 2007; Vanderpool, Chalmers, Lamb, and Whiston, 1968).

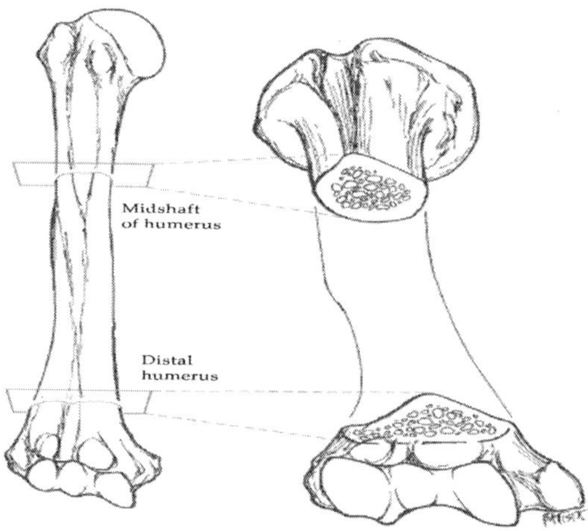

Figure 3.2. How well the distal humerus flattens anteriorly and posteriorly in a trans image of the midshaft of the humerus and the distal humerus.

Source: https://surgeryreference.aofoundation.org/orthopedic-trauma/pediatric-trauma/distal-humerus/further-reading/distal-humeral-surgical-and-developmental-anatomy.

The medial epicondyle serves as a key attachment point for the elbow's joint capsule as well as the medial (ulnar) collateral ligament, and also the forearm's superficial flexor muscles (Woods and Tullos, 1977). The posterior epicondyle is prominent at the back of the elbow, especially in flexion. It brings birth to the major collateral ligament as well as the forearm's shallow extension muscles. The radial and coronoid fossae of the humerus are shallow depressions on the distal humerus' anterior layer, immediately proximal to the capitulum and trochlea's articular faces, correspondingly. At maximal elbow flexion, such depressions enable the humerus to be near the radius and ulna. The olecranon fossa is a deep force on the distal humerus' posterior side, next to the trochlea. Whenever the elbow is stretched, the proximal portion of the ulna's olecranon line fits into this notch (Gottschalk, Eisner, and Hosalkar, 2012).

The distal humerus' articular surfaces make up the laterally 2/3rd of its distal side. The capitulum is in the laterally third, while the trochlea is in the central third. The capitulum, which is located on the anterior as well as distal parts of the humerus and therefore does not reach over the posterior surface, resembles a globe. The trochlea is a pulley-shaped layer that spans the anterior, distal, and posterior parts of the humerus, practically making a 330° circle (Helfet and Schmeling, 1993; Ring and Jupiter, 2000). The medial trochlea grows further distally as compared to the side, which explains the ulna's lateral position about the humerus. Following section, the carrying angle, which describes this orientation, is studied in more depth (Figure 3.3).

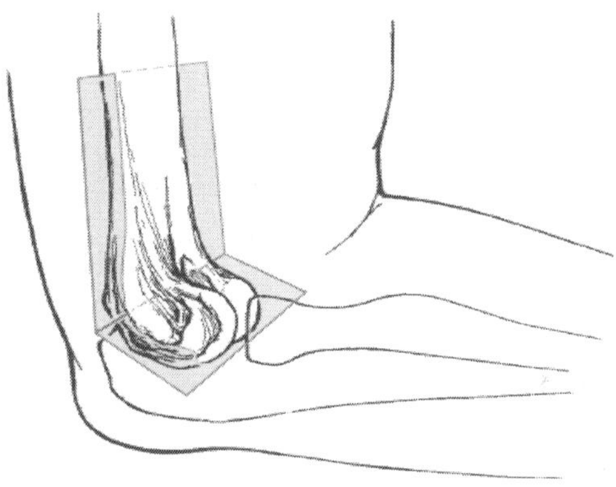

Figure 3.3. The forward curve of the humerus' distal end in a sagittal perspective.

Source: https://www.sciencedirect.com/science/article/pii/B9780128121627000114.

Hyaline cartilage covers the articular regions of the trochlea and capitulum. The capitulum cartilage thickness varies from 1.06 to 1.42 mm (0.24–0.30 mm) in 12 cadaver specimens. Subchondral bone mineralization and density seem to be most anteriorly on the capitulum, and distally as well as anteriorly on the trochlea (Frost, 1990; Jeffrey and Watt, 2003). The mineralization and density of the distal humerus imply as the distal humerus bears the most anterior and distal loads, as per Wolff's law (Cowin, 1986).

3.2.1. Proximal Ulna

The distal end of the ulna is much smaller as compared to the proximal end. The proximal ulna, the same as the distal humerus, is bent anteriorly. It is mostly composed of the olecranon as well as coronoid processes, as well as the trochlear notch that produces. This notch expresses with the humeral trochlea. The olecranon is a claw protrusion that extends anteriorly and then proximally. It is silky and readily palpable posteriorly among the two humeral epicondyles whenever the elbow is stretched. The pointy intersection among the posterior and superior faces of the olecranon feature is located distal to the two epicondyles whenever the elbow is bent, making a triangular shape along with such two bony landmarks (Pompe and Beekman, 2013; Zhang and Wang, 2018).

The olecranon is consistent along with the posterior aspect of the ulna, commonly called the ulnar crest, that is palpable over the whole distance of the ulna. The coronoid bone is located on the anterior portion of the distal ulna, and its upper side creates the trochlear notch's flooring. The distal end of the anterior side of the system is referred to as the ulnar tuberosity (Hak and Golladay, 2000; Veillette and Steinmann, 2008). On the coronoid process's outer layer lies a fine oval facet. That face, the radial notch, is where the tip of the radius articulates. Directly proximal to such a face is a fossa which serves as the muscle's point of attachment. The posterior extremity of this fossa is the supinator ridge (Figure 3.4) (Hak and Golladay, 2000).

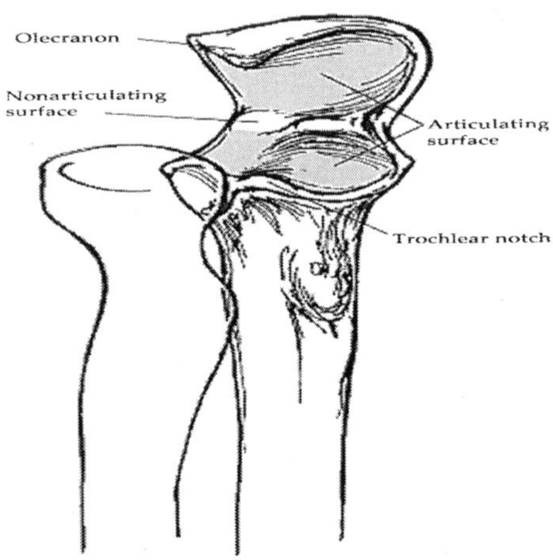

Figure 3.4. The articular surface of the trochlear groove.

Source: https://www.sciencedirect.com/topics/immunology-and-microbiology/ulna.

The trochlear notch is produced by the olecranon process's anterior side as well as the coronoid process's upper side. The trochlear notch is lined with articular cartilage and also contains a central ridge that runs proximally as well as distal throughout its length. It inserts into the innermost portion of the humeral trochlea. Anteriorly and longitudinally, the connection of the olecranon and coronoid structures in the trochlear notch is slightly reduced. The articular facet of the trochlear notch is typically split into distal and proximal joint cartilage by a nonarticular ridged region (Burton et al., 2007). The hyaline cartilage surrounding the trochlear notch is weakest crosswise as well as longitudinally and thickens through the surface's midline, with a maximum mean diameter of around 2 mm in 14 cadaver samples. Nevertheless, cartilage thickness is shown to fluctuate proximally and distally throughout the area, depending on where the articular surface of the trochlear notch is contiguous or split into discrete articular surfaces. Like in the humerus, the amount of subchondral bone development changes throughout the trochlear notch, with higher mineralization in the distal and proximal areas as compared to in the middle, indicating which the architecture of the bone is dependent on the stresses it absorbs (Figure 3.5) (Madry, van Dijk, and Mueller-Gerbl, 2010; Stewart and Kawcak, 2018).

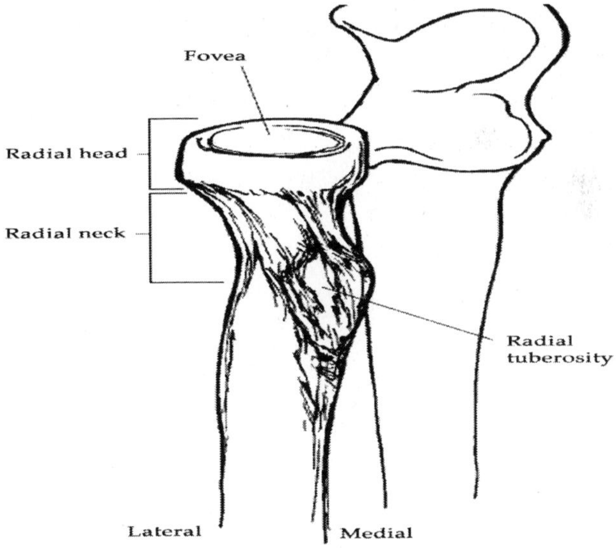

Figure 3.5. An anterior view of the proximal radius.

Source: https://musculoskeletalkey.com/fractures-of-the-ulna-and-radius/.

3.2.2. Proximal Radius

The radial head, neck, and tuberosity are all part of the proximal radius. The radial head is a disc-shaped extension of the radius at its anterior part. The fovea of the radial articulates along with the capitulum and therefore is curved on the proximal aspect of the head (Mariappan, Mohanen, and Moses, 2013). The head's rim, or peripheral aspect, is likewise articular, rotating in the ulna's radial fossa. The radial head's rim may indeed be ellipsoid or circular in a transverse perspective (YIN, 2016). The route of the distal end throughout pronation as well as supination is determined by the specific form of the rim. The rim is tallest in the middle and shallower on the sides. The distal head's rim may be felt immediately proximal to the lateral epicondyle of the humerus on the anterior face of the elbow. The particular aspect of the distal radius, such as the head as well as rim, is coated with hyaline cartilage, similar to the humerus and ulna, having widths in the foveae of cadaver samples varying from roughly 0.9 to 1.10 mm. The middle region of the fovea is said to have the deepest mineralization of the subchondral bone.

The radius's width diminishes as it moves away from the head, generating the radius's collar. The tip of the radius expands further than the circle of the head in adults, causing a constriction inside the head through which the annular ligament slips. On the medial surface of the radii, the radial tuberosity is located posterior to the radial collar. A radius's axis is somewhat bent, along with the biggest curve occurring towards the midpoint of the column, in which the pronator teres connects. The radius also may be used as a lever to change the pronator teres' lever arm (Figure 3.6) (Skie, Parent, Mudge, and Wood, 2007; Udall et al., 2009; Veeger, Kreulen, and Smeulders, 2004).

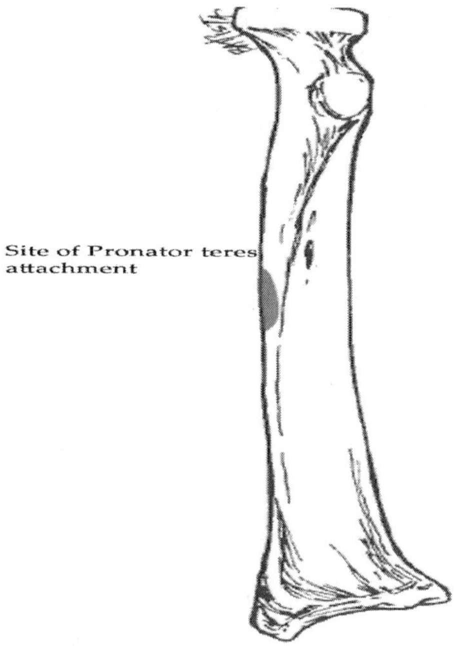

Figure 3.6. The radius in a bow form.

Source: https://journals.plos.org/plosone/article?id=10.1371/journal.pone.0258232.

By palpation, various points on the bone of the elbow may be identified (Patel and Ganley, 2012; Wilson, Ingram, Rymaszewski, and Miller, 1988). A proper physical exam must include the correct assessment of such organs. By palpation, the main bone structures may be identified:

- Medial epicondyle of the humerus;
- Lateral epicondyle of the humerus;
- Medial supracondylar ridge of the humerus;
- Lateral supracondylar ridge of the humerus;
- Olecranon process;
- Olecranon fossa of the humerus;
- Crest of the ulna;
- Radial head.

3.3. ARTICULATIONS AND SUPPORTING STRUCTURES OF THE ELBOW

Though this elbow is surrounded by a singular joint capsule, it has three separate articulations: the humeroulnar, humeroradial, as well as superior radioulnar bones. The humeroulnar and humeroradial articulations are referred to as the elbow. Nevertheless, since the superior radioulnar junction is so closely connected to another articulation, the word is occasionally used to encompass it. As a result, the doctor must be careful to determine if the elbow also relates to the superior radioulnar junction.

The explanation which follows distinguishes the appearance of humeral articulations from simply the ulna and radius. The operational contrasts between such two systems lead to that split. Flexion and extension are caused by the humeral articulations well with ulna and radius. Pronation and supination are possible because of the superior radioulnar junction (Acosta et al., 2019).

3.3.1. Humeroulnar and Humeroradial Articulations

The articulations of the humeroulnar and humeroradial bones are unique. Nevertheless, they combine to create the elbow articulation, which is a hinge joint that allows flexion and extension movement. A few of the structural supports are also shared. As a result, every unit covered with articular cartilage is detailed individually, while the supporting components for both joints are shown simultaneously. After the explanations of the joints as well as related structures, the movements permitted at the articulations are detailed simultaneously (Eckstein et al., 1995; Mason et al., 2008).

The trochlear notch of the ulna surrounds the trochlea of the humerus in the humeroulnar articulation. The opposite articular face is normally similar, only with the crest of the ulna's trochlear notch fitting snugly into the trochlear groove. Close inspection, though, shows that the fit is not flawless. The joint space ranges from 0.5 to 1.0 mm in the thickness of the trochlear notch in 15 cadaver samples carrying a load of 10 N (about 2.25 kg) and may exceed 3.0 mm medially as well as laterally. Very smaller joint gaps are recorded proximally in each of such samples. As the articular cartilage changes shape, the joint congruity enhances with higher joint stress (Figure 3.7) (Alves-Pimenta et al., 2015, 2017).

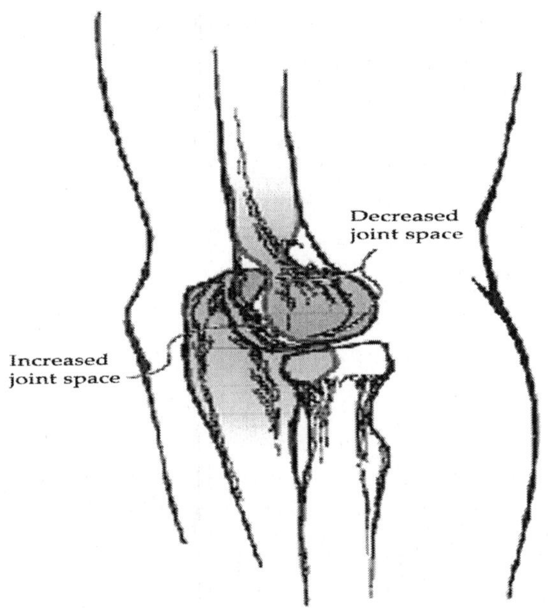

Figure 3.7. Humeroulnar articular surfaces.

Source: https://teachmeanatomy.info/upper-limb/joints/elbow-joint/.

The terminal humerus' anterior curve and the proximal ulna's corresponding curve assist establish the humeroulnar joint's respective quantities of extension and flexion mobility front twist among both joints places the articular facets in a posture that favors flexion overextension. By expanding the length, the ulna may move before all the olecranon enters the olecranon fossa, a better arrangement of such structures would result in a greater extended range of motion (ROM). The coronoid processes reaching the coronoid fossa, though, could restrict flexion early (Hulke, 1869; Morrey, 1992).

The frontal plane position of the ulna, as well as humerus, is also influenced by the geometry of its articulation. The trochlea's medial flare goes farther as compared to the lateral flare. The medial portion of the ulna's trochlear notch moves more laterally as a consequence of this growth, leading to a medial displacement of the ulna regarding the humerus. Whereas the carrying tilt is often used to define this position, valgus is a more general phrase. A valgus deviation is a medial deflection of a distal section relative to the proximal (Thomas and Barrington, 2003). Varus is the polar opposite, a medial deviation of a distal section relative to the proximal. Whenever the among the distal and proximal sections is 180° (commonly characterized as 0°), the joint is in a stable position between varus and valgus. The valgus alignment, also known as the carrying angle, of the elbow, has been the subject of much research. Carrying angles of 10–15° have been documented. Though many textbooks claim that women have a larger carrying angle as compared to the males, rigorous measurements revealed that there is no substantial distinction in carrying angles among the sexes (Bevernage and Leemrijse, 2009; Julian, 1984).

The radius head rests on the capitulum of the humerus in the humeroradial articulation. Whenever the elbow is stretched, the head of the radius articulates only with a piece of the capitulum since it is located on the front side of the distal humerus. During elbow flexion, the interaction between the humerus and the radius improves (Alvarez, Patel, Nimberg, and Pearlman, 1975; Fioretta, Rotolo, and Zanasi, 1984; Lansinger and Måre, 1981).

3.3.2. Superior Radioulnar Joint

Although being enclosed inside the capsule of the elbow junction, the upper radioulnar joint is functionally separate from humeral articulations. The articulation is characterized as only one pivot joint. Dissimilar to proximal articulations, the upper radioulnar joint's bony architecture, which consists of the lip of the posterior aspect as well as the radial aspect on the ulna, gives almost no stability to the junction. Thus, the upper radioulnar junction is supported by the fibrous tissue around it, such as the capsules and LCL, the annular ligament, the interosseous membrane, and the oblique cord. The capsule and LCL have indeed been defined, thus no more examination is necessary. The structure and composition of the other ligaments are described below.

The annulus ligament is a thick, fibrous ring that encircles the neck of the radius and attaches to the upper and lower edges of the radial notch on the ulna. Therefore, it creates a loop around the radii, including its major connections on the ulna and several weak connections to the capsule, posterior portion of the trochlea, and radial neck. The inner layer of the annulus ligament is covered by fibrocartilage, which provides further rigidity and elasticity (Bozkurt et al., 2005; Roofe, 1940). The greater mechanical rigidity is crucial because, unlike other ligaments, that link to the joints they sustain, the annular ligament largely works as a sling, preventing the radius from slipping. The impact on the usual mobility of the upper radioulnar joint is minimal or nonexistent.

The annular ligament connects the radii to the ulna, preventing lateral subluxation effectively. Additionally, the annular ligament is the principal barrier to subluxation or displacement of the upper radioulnar junction at its distal end. Usually, this injury is caused by pressure that pulls the forearm laterally from the elbow, including when carrying or hanging a kid by the arms (Dik, Van Den Belt, and Keg, 1991).

3.4. MECHANICS AND PATHOMECHANICS OF MUSCLE ACTIVITY AT THE ELBOW

3.4.1. Elbow Flexor Muscles

The biceps brachii, brachialis, and brachioradialis are the major flexors of the elbow. Its category also includes the pronator teres, which help with dynamic elbow flexion. The acts that almost every muscle is responsible for are listed here (Jakubowicz and Malinowski, 1992). The known information on every muscle's involvement in the concerted action of elbow flexion is provided as every muscle is studied separately. This learning is dependent on both electromyographic (EMG) data and numerical simulations of the area (Figure 3.8) (Naito, 2004).

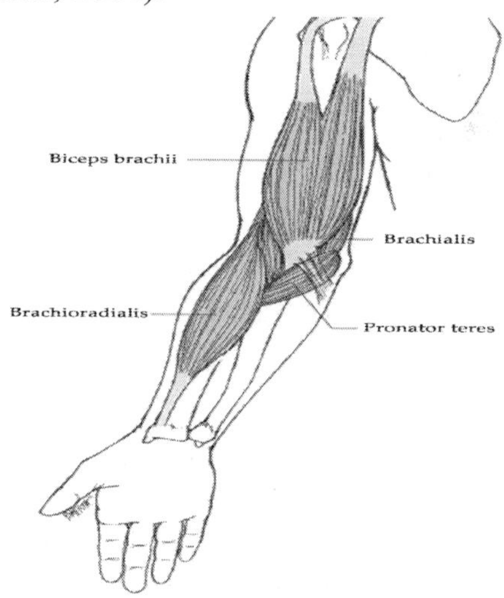

Figure 3.8. The biceps brachii, brachialis, brachioradialis, and pronator teres are the major flexion muscles of the forearm.

Source: https://www.sportsinjuryclinic.net/sport-injuries/elbow-pain/elbow-joint-muscles.

The biceps brachii is a two-headed fusiform muscle. Its connections cover the elbow and shoulder, and it spans the humeral and radioulnar articulations at the elbow. The glenohumeral, humeroulnar, and humeroradial articulations, and the upper radioulnar joint, are all affected by biceps brachii contractions.

The biceps brachii flexes the elbow as well as supinates the biceps without a doubt (Fellows and Rack, 1987; Jakubowicz and Malinowski, 1992). However, both head of the biceps brachii is involved in such activities; their relative importance is unknown. In just about all participants, according to Basmajian and De Luca, the posterior aspect is more responsive as compared to the short head throughout concentric elbow flexion with unresisted supination. Following elbow flexion, Stewart et al. found no change in activity between the two heads, independent of elbow location or contractile speed. When contrasted to the anterior surface, research of 10 patients with long-term (average of 3.2 years) bursting of the tendon of the large head of the biceps found power impairments of 10–15% for elbow flexion as well as lower as compared to the 2% for supination (Marcolin et al., 2018).

According to EMG data, based on the forearm posture, the degree of resistance at the time of motion, and the velocity of the movement, the muscles of the forearm have unique separate functions in elbow movement. Such factors influence the biceps brachii's function throughout elbow movement. After the explanation of the remaining elbow flexors, the precise situations wherein the forearm engages in elbow flexion as well as forearm supination is examined.

Biceps brachii is often referred to be a shoulder flexor (Boyd and Anderson, 1961; MacDougall, Sale, Alway, and Sutton, 1984). EMG evidence from Basmajian and De Luca supports this widely believed belief. In shoulder flexion, either head of the biceps brachii is active, although the long head is much more active in most people. According to the research on five cadaver shoulders, the small head of the biceps brachii has a significant lever arm for shoulder flexion, but the posterior aspect has none.

Since this biceps brachii spans either the shoulder as well as the elbow, variations in position at each joint alter muscle mass. Latent elbow flexion shortens the muscle, causing it to become slack. A similar is true for mild shoulder flexion. Involuntary elbow, as well as shoulder flexion, expand the muscle by putting it in a longer posture (Warner and McMahon, 1995).

According to the length-tension connection, a muscle's force of contract increases as it is extended, whereas its contraction force decreases as it is reduced. (That relationship's specifics are revealed.) Isolated contract of the biceps brachii generates elbow and shoulder flexion at the same time since it is a two-joint muscle. If enough elbow, as well as shoulder flexion, occurs at the same time, the biceps brachii might well be reduced to the point where it could produce minimal force. Recurrent inadequacy is the term for this condition. With elbow flexion, however, shoulder extension elongates the biceps brachii thus raising the biceps contractility.

Elbow flexion and shoulder extension are produced by contracting the biceps brachii muscle while simultaneously contracting a shoulder extensor muscle, allowing the biceps brachii muscle to retain its length. Throughout a maximal voluntary contraction (MVC) of the elbow flexors, Allen et al. observe minor shoulder hyperextension. According to such researchers, extending the biceps in line with the length-tension relation of a muscle increases the contraction power. Clinicians may exploit this positional impact to help patients improve their elbow flexion power (Rayan, Jensen, and Duke, 1992; Youm et al., 1979).

The biceps brachii has been described as a shoulder abductor in several studies. In the context of long-term bursting of the tendon of the elongated head of the biceps brachii, Sturzenegger et al. (1986) indicate an estimated 8% loss in shoulder abduction power. According to an examination of five cadaver shoulders, all heads of the biceps brachii contain abduction moment arms, showing as the muscle may produce shoulder abduction. Nevertheless, no investigations on the Emg of the biceps brachii in shoulder abduction are available. The cadaver research also reveals as the biceps' long head may generate a sideways rotation motion. EMG findings, on the other hand, reveal no biceps brachii activity in lateral rotation with occasional short head action in medial rotation (Moorman III, Silver, Potter, and Warren, 1996). These findings back up the theory that the biceps brachii helps in shoulder abduction and twisting. To define their responsibilities, further study is required. The long head of the biceps brachii has been identified by several writers as a key dynamic stabilizer of the glenohumeral joint. The upper end of the elongated head of the biceps brachii tendon runs practically similar to the supraspinatus muscle so operates, in the same way, to squeeze the glenohumeral joint to support it.

Based on the joint's rotation, a careful analysis of cadavers reveals here that the biceps may offer substantial prevention mostly against anterior and posterior displacement of the glenohumeral joint. Other cadaver investigations (Chansky and Iannotti, 1991; Singh et al., 2020) confirm the biceps' participation in glenohumeral joint stabilization in the anterior-posterior (AP) and superior-inferior (SI) orientations. In people with no shoulder disease, though, EMG data demonstrate little activation of the biceps brachii to support the glenohumeral joint under forces that sublux the joint inferiorly (Sharma et al., 2022).

Even though such studies seem to oppose each other, it is vital to acknowledge the difficulties in analyzing them. The cadaver investigations show that the biceps brachii can support the glenohumeral joint. The EMG data comes from research on people having normal joint stability who only moved their humerus in the lower position. The involvement of the biceps brachii in maintaining the glenohumeral joint within every orientation in living persons with stable shoulders requires further research.

The involvement of the biceps brachii in supporting the glenohumeral joint in certain orientations is unknown till such data exists. While other adjusters are weakened, though, there is considerable proof to substantiate its function as a glenohumeral joint adjuster (Hess, 2000; Lippitt and Matsen, 1993).

3.4.1.1. Effects of Weakness

Biceps brachii weakening leads to a decrease of energy in elbow flexion and supination. Nevertheless, a report of a patient including a unilateral lesion of the musculocutaneous nerve and full denervation of the biceps brachii shows a patient with outstanding function due to the compensations supplied through other elbow muscles (Bhattarai and Poudel, 2009; de Mendonça, Gepp, and Correa, 2016). Noteworthy is that neither "good function" nor power metrics are specified in the study. Although the elbow includes multiple flexor muscles, the biceps brachii is the major flexor, as well as its weakening, leads to a significant drop in power. Nevertheless, the residual elbow flexor muscle seems to retain a substantial amount of function. Likewise, biceps brachii weakening results in a large drop in supination ability, while the residual muscles which supinate the elbow minimize the operational deficit (Xu et al., 2005).

In addition to stiffness in shoulder flexion, biceps brachii dysfunction may appear as minor difficulty in shoulder extension. Nevertheless, since the major shoulder flexors are so massive and powerful, isolated biceps weakening rarely results in an operationally meaningful decrease of shoulder flexion ability. Nonetheless, diminished biceps brachii power in the context of rotator cuff disease can exacerbate glenohumeral joint dislocation.

3.4.1.2. Effects of Tightness

Stiffness of the biceps brachii muscle can reduce elbow expansion, pronation, and possibly shoulder enhanced version range of motion (ROM). As a biarticular, or two-joint, muscle, though, the consequences of tension at one contact are modified by the location of the second contact. The interaction among elbow and shoulder locations and the impact of this interplay on the biceps brachii can aid the doctor in identifying structures that could restrict the extension range of motion of the elbow joint (Bailey, Samuel, Warner, and Stokes, 2013; Chen et al., 2017).

3.5. COMPARISONS AMONG THE ELBOW FLEXORS

Previously, we addressed the dimensions of muscular strength, energy output, and motion creation. It also examines the aspects which affect muscular function, such as muscle growth, degree of activation, and recruitment level. Such characteristics differentiate the elbow flexors from each other. Even though there are four basic elbow flexors, each seems to contribute uniquely to the functioning of the elbow (Knapik, Jones, Meredith, and Evans, 1987). Throughout this part, the structural features of such muscles, including their physiological cross-sectional areas (PCSOs), moment arms, as well as lengths, are compared. The role which each muscle seems to have in the functioning of the elbow is then addressed in light of EMG data (Rudroff, Staudenmann, and Enoka, 2008).

3.5.1. Structural Comparison of Elbow Flexors

PCSA is a measurement of the quantity and diameter of accessible muscle fibers in a muscle, and hence an indicator of the muscle's force-producing capability. The bigger the PCSA, the larger the force output capability. The PCSA of the brachialis (about 5.5–8.0 cm^2) is the biggest. The biceps brachii is the next one (about 4.5 cm^2), preceded by the pronator teres (roughly 4.5 cm^2) (estimated 4.0 cm^2). The lowest PCSA (about 1.3 cm^2) is found in the brachioradialis (Boland, Spigelman, and Uhl, 2008).

A muscle's mechanical value is measured by its moment arm and muscle length, in combination with the contractility. A muscle's moment arm (M r F) is a combination of its contraction force (F) and its moment arm (M r F) (r). The brachioradialis has the biggest moment arm, connecting axially on the radius, preceded by the biceps brachii, and finally the brachialis. The moment arm of the pronator teres is the shortest (Figure 3.9).

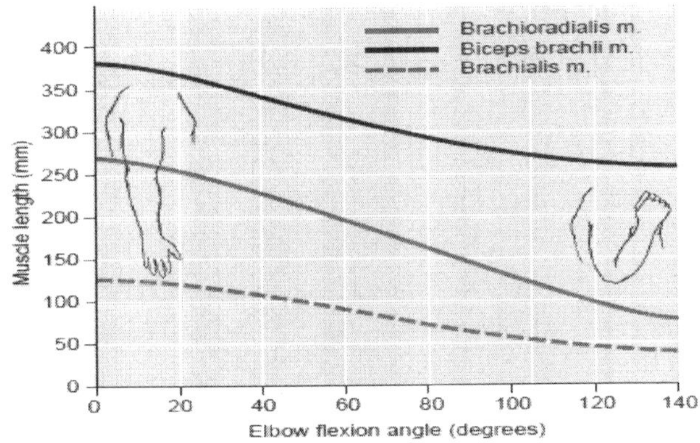

Figure 3.9. The widths of the brachioradialis, biceps brachii, and brachialis muscles in contrast.

Source: https://www.osmosis.org/answers/brachialis.

Whereas the brachialis is the greatest elbow flexor, its moment arm puts it at a mechanical obstacle. It is hard to know how much every muscle contributes to the overall flexion moment delivered to the elbow. Nevertheless, biomechanical balanced that the brachialis and biceps brachii contribute the most to elbow flexion torque when the forearm is in neutral, while their relative proportions are still debated. The proportionate involvement of the elbow flexor muscles changes throughout elbow flexion, according to models (Lai, Krishna, and Pelly, 1981).

Such muscles' variable inputs to the total moment may be characterized partly by the moment arms, that change with the joint excursion. The moment arms of the elbow flexors alter dramatically across the span of extension and flexion, or during pronation and supination, according to anatomical investigations and computer simulations. In the second period of the span of elbow flexion, the applying angles approach their maximum. At 90° and 110° of elbow flexion, the biceps brachii achieves its maximal moment arm. At 90° of flexion, the brachialis moment arm also peaks. The pronator teres seems to peak sooner, at 75° while the brachioradialis achieves its highest moment arm from 100° to 120° of elbow flexion. As the elbow travels over its range of motion, the estimated moment arms of such four muscles.

Pigeon and Feldman (1996) and Murray et al. (2005) provided the information for the study. The shift in the moment arm with full extension to full flexion is estimated to be between 30% and 85%. A muscular contraction generates a moment that is related to the muscle's moment arm. Whenever a muscle's moment arm is long, it creates a bigger moment for a given amount of contract. As a result of the varying moment arms, the elbow flexors' ability to create a moment change considerably over the span.

Every elbow flexor's length varies dramatically across the range of motion, rising as the elbow is elongated and reducing as the elbow is flexed. A muscle's capacity to create force increases as it is stretched and decreases as it is reduced, per the length-tension connection. As a result, whenever the length-tension of the elbow is stretched, the elbow flexors extend, making force generation easier.

The figure depicts estimations of muscle sizes of the elbow flexor muscles over the range of motion. In the expanded posture, though, the flexors' moment arms are rather tiny, reducing the muscles' ability to create torque. As a result, the influence of elbow joint location on elbow flexor length differs significantly from the impact on muscle moment arm. The location where the whole elbow flexors produce the maximum flexion torque would be in the middle of flexion, since such two major parameters determining muscle function, moment arm and muscle length, change considerably throughout elbow motion. Figure 3.8 depicts the overall association between elbow flexion strength and elbow flexion location (Knapik, Fausett, Gilmore, and Liu, 2017; Williams and Stutzman, 1959). Although neither moment arm nor the muscle length is ideal at the mid-position (Figure 3.8). Instead, the largest elbow flexion torque production is achieved by striking a balance among muscle length and moment arm. Peak elbow flexion isometric power is known to cause when the elbow is bent to 90. Such a result, nevertheless, is collected mainly in 25–30 steps across the elbow range of motion. According to the other research, maximal force production in females during isometric as well as isokinetic contracts and in males throughout isokinetic contractions occurred at 70 of elbow flexion. According to this research, men's isometric contraction force peaks at 90°, although the exact region of the peak force differs widely across respondents. Regardless of such differences, the basic fact is clear: elbow flexion ability peaks or something in the center of elbow flexion range of motion, when neither the moment arms nor the lengths of the muscle are optimum for force generation.

The quantity of excursion induced by a stretch is also influenced by the extent of a muscle's moment arm and fiber length. As previously mentioned, muscles with a small moment arm may induce a lot of joint excursions for a little degree of shortening, whereas a muscle with a long moment arm causes lesser joint excursion for a similar amount of contraction. As a result, the brachialis with the smallest moment arm is well-suited to moving the elbow across a wide range of motion. Long muscular fibers in the biceps and brachioradialis assist in the capacity to effectively move the elbow over its complete range of motion.

Finally, the elbow flexor architecture shows here that brachialis and biceps brachii are most adapted to create substantial flexor moments at the elbow. Architectural characteristics in the elbow flexors promote dynamic elbow mobility along its complete arc. While knowing every muscle's output potential is useful, a complete comprehension of the significance of the elbow flexors necessitates an awareness of the info collected on the recruitment sequence during elbow movement.

3.5.2. Comparisons of Flexor Muscle Activity

Just several managed EMG types of research have indicated the personal donations of every one of the elbow flexors to movement. Some of the research mentioned by Basmajian (1962) and De Luca, Roy, and Erim (1993) are considered classics. They trying to follow is a concise summary of their findings. According to the data presented, the biceps brachii, brachialis, and brachioradialis operate cohesively, with every muscle performing its function. Even so, the writers also highlight that the actions of such muscles vary significantly amongst the healthy population. Despite this limitation, the data indicate usable trends for every muscle (Khemlani, Carr, and Crosbie, 1999).

The biceps brachii is stimulated in the majority of other people all through supinated elbow flexion, irrespective of flexion speed or rigidity. Whenever the forearm is partly pronated, just resistance has seemed to recruit the biceps brachii. Whenever the forearm is completely pronated, the muscle is only recruited while resistance is applied, and anyway, it would seem to be only partly enabled. Likewise, the biceps are involved throughout forearm supination with resistance whenever the elbow is flexed. Whenever the elbow is stretched, even so, just strenuously trying to resist supination stimulates the biceps brachii, which is frequently accompanied by mild elbow flexion (Li et al., 2022).

The biceps brachii is most operative whenever the elbow is trying to move in both flexion and supination, according to such findings. That those are the movements largely linked to the biceps brachii, so this conclusion is quite rational. Because a muscle's role is to trim down and the resulting motion is a function of the muscle's line of pull, it follows that a muscle is most active when performing the motions for which it was designed.

Until one of these movements is unwanted, the options are to recruit a muscle whose exact action is desired or to block the muscle with several actions whilst stabilizing the joint in an intended way with some other muscle. This second option seems to function throughout both severely opposed elbow flexion with the forearm pronated and resisted supination of the forearm with the elbow straightened. Within every instance, the biceps help the intended movement while simultaneously adding a force in a reverse way. The biceps brachii is restricted from fullest capacity during resisted elbow flexion with pronation so that its supination function does not interact with the pronation posture. Resistant supination with elbow flexion likewise suppresses biceps brachii activity to prevent elbow extension from interfering. In both instances, the biceps brachii is partly recruited.

The brachialis, on the other hand, is active anytime the elbow is flexed, independent of forearm location, opposition, or motion. The brachialis is connected to the ulna, which rotates minimally during pronation and supination of the forearm. Therefore, the brachialis is unchanged by the location of the forearm. Its activation is effective, as it possesses the most PCSA and, hence, the greatest potential contractile power. As a result of the muscle's small moment arm, it can also drive the elbow throughout its whole range of motion. The brachialis' continuous activity throughout elbow flexion has earned it the moniker "workhorse" of elbow flexion (Murray, Buchanan, and Delp, 2000).

The brachioradialis seems to be engaged during elbow flexion with resistance, especially in the slightly pronated and pronated postures, as well as in the supinated posture. Additionally, it is activated with fast elbow flexion. This last behavior is rather unexpected due to the muscle's lengthy moment arm. It is theorized, though, that the placement of the muscle throughout the length of the forearm offers a significant unifying influence for the elbow against the centrifugal motion that tends to distract the elbow during fast movements. As mentioned before, the pronator teres seems to only be engaged during elbow flexion or pronation when resistance is applied.

The brachialis and brachioradialis are forcefully stimulated by resistive elbow flexion with the forearm pronated, whereas the biceps are only partly activated, confirming the concept that elbow flexion with the elbow pronated is weak. Try comparing the complexity of chin-up exercises done with the elbow supinated and pronated to verify this notion. No research has fully replicated the previous trials. Nevertheless, many investigations are inconclusive. At elbow flexion, Stewart et al. also found brachialis action irrespective of forearm position. In comparison to supination, biceps brachii action is reduced and brachioradialis activity is higher in elbow flexion with elbow pronated. However, the brachioradialis muscle is functional with no resistance in such studies. According to the other study, whenever the biceps brachii is entirely recruited at MVC, the brachioradialis is not entirely engaged, implying that the brachioradialis serves as a reservoir for additional power. The moderately pronated grip generated more power than the pronated posture in a test of very well professional rowers. Considering the absence of actual EMG data, that study adds to the growing body of evidence indicating the forearm flexors are more fully involved when the elbow is only partially pronated as compared to when it is fully pronated (Lowery, Nolan, and O'malley, 2002; Tang and Rymer, 1981).

The results back with Basmajian and De Luca's original claim that there is exquisite synchronization between elbow flexors in elbow and forearm motion. Knowing how many muscles connect allows the therapist to do a thorough evaluation of the muscles that flex the elbow to develop an intervention strategy by focusing treatment on the specific issue.

REFERENCES

1. Acosta, B. J., Cerezal, L., López, P. M. D., Alba, B., Resano, S., & Blázquez, S. J., (2019). The elbow: Review of anatomy and common collateral ligament complex pathology using MRI. *Insights into Imaging, 10*(1), 1–25.
2. Alvarez, E., Patel, M., Nimberg, G., & Pearlman, H., (1975). Fracture of the capitulum humeri. *The Journal of Bone and Joint Surgery; American, 57*(8), 1093–1096.
3. Alves-Pimenta, S., Ginja, M. M., Colaço, J., Fernandes, A. M., Melo-Pinto, P., & Colaço, B., (2015). Curvature radius measurements from the ulnar trochlear notch in large dogs. *The Anatomical Record, 298*(10), 1748–1753.
4. Alves-Pimenta, S., Ginja, M. M., Fernandes, A. M., Ferreira, A. J., Melo-Pinto, P., & Colaço, B., (2017). Computed tomography and radiographic assessment of congruity between the ulnar trochlear notch and humeral trochlea in large breed dogs. *Veterinary and Comparative Orthopedics and Traumatology, 30*(01), 8–14.
5. Anglen, J., (2005). Distal humerus fractures. *JAAOS-Journal of the American Academy of Orthopedic Surgeons, 13*(5), 291–297.
6. Bailey, L., Samuel, D., Warner, M., & Stokes, M., (2013). Parameters representing muscle tone, elasticity and stiffness of biceps brachii in healthy older males: Symmetry and within-session reliability using the MyotonPRO. *Journal of Neurological Disorders, 1*(1), 1–7.
7. Basmajian, J. V., (1962). Muscles alive. Their functions revealed by electromyography. *Academic Medicine, 37*(8), 802.
8. Bevernage, B. D., & Leemrijse, T., (2009). Hallux varus: Classification and treatment. *Foot and Ankle Clinics, 14*(1), 51–65.
9. Bhattarai, C., & Poudel, P., (2009). Unusual variation in musculocutaneous nerves. *Kathmandu University Medical Journal, 7*(4), 408–410.
10. Boland, M. R., Spigelman, T., & Uhl, T. L., (2008). The function of brachioradialis. *The Journal of Hand Surgery, 33*(10), 1853–1859.
11. Bowden, R. E., (1983). Anatomy, radiology and Kinesiology of hand-unit. A biokinetic and functional unit composed of distal half of arm, elbow, forearm, wrist, palm and digits. *Journal of Anatomy, 136*(3), 596.

12. Boyd, H., & Anderson, L., (1961). A method for reinsertion of the distal biceps brachii tendon. *JBJS, 43*(7), 1041–1043.
13. Bozkurt, M., Acar, H. İ., Apaydin, N., Leblebicioglu, G., Elhan, A., Tekdemir, İ., & Tonuk, E., (2005). The annular ligament: An anatomical study. *The American Journal of Sports Medicine, 33*(1), 114–118.
14. Bunata, R. E., Brown, D. S., & Capelo, R., (2007). Anatomic factors related to the cause of tennis elbow. *JBJS, 89*(9), 1955–1963.
15. Burkhart, K. J., Nijs, S., Mattyasovszky, S. G., Wouters, R., Gruszka, D., Nowak, T. E., & Müller, L. P., (2011). Distal humerus hemiarthroplasty of the elbow for comminuted distal humeral fractures in the elderly patient. *Journal of Trauma and Acute Care Surgery, 71*(3), 635–642.
16. Burton, N., Comerford, E., Bailey, M., Pead, M., & Owen, M., (2007). Digital analysis of ulnar trochlear notch sclerosis in Labrador retrievers. *Journal of Small Animal Practice, 48*(4), 220–224.
17. Callaway, G., Field, L., Deng, X. H., Torzilli, P., O'brien, S., Altchek, D., & Warren, R., (1997). Biomechanical evaluation of the medial collateral ligament of the elbow. *JBJS, 79*(8), 1223–1231.
18. Chansky, H. A., & Iannotti, J. P., (1991). The vascularity of the rotator cuff. *Clinics in Sports Medicine, 10*(4), 807–822.
19. Chen, J., O'Dell, M., He, W., Du, L. J., Li, P. C., & Gao, J., (2017). Ultrasound shear wave elastography in the assessment of passive biceps brachii muscle stiffness: Influences of sex and elbow position. *Clinical Imaging, 45*, 26–29.
20. Cowin, S. C., (1986). *Wolff's Law of Trabecular Architecture at Remodeling Equilibrium*.(pp. 4-7)
21. Dalton, B., Jakobi, J., Allman, B., & Rice, C., (2010). Differential age-related changes in motor unit properties between elbow flexors and extensors. *Acta Physiologica, 200*(1), 45–55.
22. De Luca, C. J., Roy, A. M., & Erim, Z., (1993). Synchronization of motor-unit firings in several human muscles. *Journal of Neurophysiology, 70*(5), 2010–2023.
23. De Mendonça, C. M., Gepp, R., & Correa, J. F. G., (2016). Outcome following phrenic nerve transfer to musculocutaneous nerve in patients with traumatic brachial palsy: A qualitative systematic review. *Acta Neurochirurgica, 158*(9), 1793–1800.

24. Dik, K., Van, D. B. A., & Keg, P., (1991). Ultrasonographic evaluation of fetlock annular ligament constriction in the horse. *Equine Veterinary Journal, 23*(4), 285–288.
25. Eckstein, F., Müller-Gerbl, M., Steinlechner, M., Kierse, R., & Putz, R., (1995). Subchondral bone density in the human elbow assessed by computed tomography osteoabsorptiometry: A reflection of the loading history of the joint surfaces. *Journal of Orthopedic Research, 13*(2), 268–278.
26. Fellows, S., & Rack, P., (1987). Changes in the length of the human biceps brachii muscle during elbow movements. *The Journal of Physiology, 383*(1), 405–412.
27. Fioretta, G., Rotolo, F., & Zanasi, L., (1984). The treatment of fractures of the capitulum of the humerus. *Italian Journal of Orthopedics and Traumatology, 10*(1), 81–84.
28. Frost, H. M., (1990). Skeletal structural adaptations to mechanical usage (SATMU): 3. The hyaline cartilage modeling problem. *The Anatomical Record, 226*(4), 423–432.
29. Fukunaga, T., Kawakami, Y., Kuno, S., Funato, K., & Fukashiro, S., (1997). Muscle architecture and function in humans. *Journal of Biomechanics, 30*(5), 457–463.
30. Garland, S., Enoka, R., Serrano, L., & Robinson, G., (1994). Behavior of motor units in human biceps brachii during a submaximal fatiguing contraction. *Journal of Applied Physiology, 76*(6), 2411–2419.
31. Gottschalk, H. P., Eisner, E., & Hosalkar, H. S., (2012). Medial epicondyle fractures in the pediatric population. *JAAOS-Journal of the American Academy of Orthopedic Surgeons, 20*(4), 223–232.
32. Hak, D. J., & Golladay, G. J., (2000). Olecranon fractures: Treatment options. *JAAOS-Journal of the American Academy of Orthopedic Surgeons, 8*(4), 266–275.
33. Helfet, D. L., & Schmeling, G. J., (1993). Bicondylar intraarticular fractures of the distal humerus in adults. *Clinical Orthopedics and Related Research,* (292), 26–36.
34. Hess, S., (2000). Functional stability of the glenohumeral joint. *Manual Therapy, 5*(2), 63–71.
35. Hulke, J., (1869). Note on a large saurian humerus from the Kimmeridge clay of the Dorset coast. *Quarterly Journal of the Geological Society, 25*(1, 2), 386–389.

36. Jakubowicz, M., & Malinowski, A., (1992). Metric features of the human biceps brachii, brachialis and brachioradialis muscles. *Folia Morphological, 51*(1).

37. Jeffrey, D., & Watt, I., (2003). Imaging hyaline cartilage. *The BRITISH Journal of Radiology, 76*(911), 777–787.

38. Julian, R., (1984). Valgus-varus deformity of the intertarsal joint in broiler chickens. *The Canadian Veterinary Journal, 25*(6), 254.

39. Keskula, D. R., (2004). Kinesiology: The mechanics and pathomechanics of human movement. *Physical Therapy, 84*(3), 291.

40. Khemlani, M., Carr, J., & Crosbie, W., (1999). Muscle synergies and joint linkages in sit-to-stand under two initial foot positions. *Clinical Biomechanics, 14*(4), 236–246.

41. Knapik, D. M., Fausett, C. L., Gilmore, A., & Liu, R. W., (2017). Outcomes of nonoperative pediatric medial humeral epicondyle fractures with and without associated elbow dislocation. *Journal of Pediatric Orthopedics, 37*(4), e224–e228.

42. Knapik, J. J., Jones, B. H., Meredith, C., & Evans, W. J., (1987). Influence of a 3.5 day fast on physical performance. *European Journal of Applied Physiology and Occupational Physiology, 56*(4), 428–432.

43. Lai, M., Krishna, B., & Pelly, A., (1981). The brachioradialis myocutaneous flap. *British Journal of Plastic Surgery, 34*(4), 431–434.

44. Lansinger, O., & Måre, K., (1981). Fracture of the capitulum humeri. *Acta Orthopedic Scandinavica, 52*(1), 39–44.

45. Li, G., Wang, M., Arippa, F., Barr, A., Rempel, D., Liu, Y., & Harris, A. C., (2022). Professional and high-level gamers: Differences in performance, muscle activity, and hand kinematics for different mice. *International Journal of Human-Computer Interaction, 38*(8), 691–706.

46. Lieber, R. L., Murray, W. M., Clark, D. L., Hentz, V. R., & Fridén, J., (2005). Biomechanical properties of the brachioradialis muscle: Implications for surgical tendon transfer. *The Journal of Hand Surgery, 30*(2), 273–282.

47. Lippitt, S., & Matsen, F., (1993). Mechanisms of glenohumeral joint stability. *Clinical Orthopedics and Related Research,* (291), 20–28.

48. Lowery, M., Nolan, P., & O'malley, M., (2002). Electromyogram median frequency, spectral compression and muscle fiber conduction

velocity during sustained sub-maximal contraction of the brachioradialis muscle. *Journal of Electromyography and Kinesiology, 12*(2), 111–118.
49. MacDougall, J., Sale, D., Alway, S., & Sutton, J., (1984). Muscle fiber number in biceps brachii in bodybuilders and control subjects. *Journal of Applied Physiology, 57*(5), 1399–1403.
50. Madry, H., Van, D. C. N., & Mueller-Gerbl, M., (2010). The basic science of the subchondral bone. *Knee Surgery, Sports Traumatology, Arthroscopy, 18*(4), 419–433.
51. Marcolin, G., Panizzolo, F. A., Petrone, N., Moro, T., Grigoletto, D., Piccolo, D., & Paoli, A., (2018). Differences in electromyographic activity of biceps brachii and brachioradialis while performing three variants of curl. *PeerJ, 6*, e5165.
52. Mariappan, E., Mohanen, P., & Moses, J., (2013). A newer technique of distal ulna reconstruction using proximal fibula and TFCC reconstruction using palmaris longus tendon following wide resection of giant cell tumor of distal ulna. *Case Reports in Orthopedics, 2013.*
53. Mason, D. R., Schulz, K. S., Fujita, Y., Kass, P. H., & Stover, S. M., (2008). Measurement of humeroradial and humeroulnar transarticular joint forces in the canine elbow joint after humeral wedge and humeral slide osteotomies. *Veterinary Surgery, 37*(1), 63–70.
54. Moorman, III. C. T., Silver, S. G., Potter, H. G., & Warren, R. F., (1996). Proximal rupture of the biceps brachii with slingshot displacement into the forearm. A case report. *JBJS, 78*(11), 1749–1752.
55. Morrey, B., (1992). Primary degenerative arthritis of the elbow. Treatment by ulnohumeral arthroplasty. *The Journal of Bone and Joint Surgery; British, 74*(3), 409–413.
56. Murray, W. M., Buchanan, T. S., & Delp, S. L., (2000). The isometric functional capacity of muscles that cross the elbow. *Journal of Biomechanics, 33*(8), 943–952.
57. Naito, A., (2004). Electrophysiological studies of muscles in the human upper limb: The biceps brachii. *Anatomical Science International, 79*(1), 11–20.
58. Patel, N. M., & Ganley, T. J., (2012). Medial epicondyle fractures of the humerus: How to evaluate and when to operate. *Journal of Pediatric Orthopedics, 32*, S10–S13.

59. Pigeon, P., & Feldman, A. G., (1996). Moment arms and lengths of human upper limb muscles as functions of joint angles. *Journal of Biomechanics, 29*(10), 1365–1370.
60. Polatsch, D. B., Melone, Jr. C. P., Beldner, S., & Incorvaia, A., (2007). Ulnar nerve anatomy. *Hand Clinics, 23*(3), 283–289.
61. Pompe, S. M., & Beekman, R., (2013). Which ultrasonographic measure has the upper hand in ulnar neuropathy at the elbow? *Clinical Neurophysiology, 124*(1), 190–196.
62. Rayan, G. M., Jensen, C., & Duke, J., (1992). Elbow flexion test in the normal population. *The Journal of Hand Surgery, 17*(1), 86–89.
63. Ring, D., & Jupiter, J. B., (2000). Fractures of the distal humerus. *Orthopedic Clinics, 31*(1), 103–113.
64. Roofe, P. G., (1940). Innervation of annulus fibrosus and posterior longitudinal ligament: Fourth and fifth lumbar level. *Archives of Neurology & Psychiatry, 44*(1), 100–103.
65. Rudroff, T., Staudenmann, D., & Enoka, R. M., (2008). Electromyographic measures of muscle activation and changes in muscle architecture of human elbow flexors during fatiguing contractions. *Journal of Applied Physiology, 104*(6), 1720–1726.
66. Sharma, D., Rubel, K. E., Ye, M. J., Campiti, V. J., Ting, J. Y., Burgin, S. J., & Illing, E. A., (2022). Response to: "Cadaver surgical education and research under the SARS-CoV-2 pandemic in Japan". *Otolaryngology-Head and Neck Surgery, 166*(5), 1004–1004.
67. Shiba, R., Sorbie, C., Siu, D. W., Bryant, J. T., Cooke, T. D. V., & Wevers, H. W., (1988). Geometry of the humeroulnar joint. *Journal of Orthopedic Research, 6*(6), 897–906.
68. Singh, R. P., Pandey, P. M., Mridha, A. R., & Joshi, T., (2020). Experimental investigations and statistical modeling of cutting force and torque in rotary ultrasonic bone drilling of human cadaver bone. *Proceedings of the Institution of Mechanical Engineers, Part H: Journal of Engineering in Medicine, 234*(2), 148–162.
69. Skie, M. C., Parent, T. E., Mudge, K. M., & Wood, V. E., (2007). Functional deficit after transfer of the pronator teres for acquired radial nerve palsy. *The Journal of Hand Surgery, 32*(4), 526–530.
70. Solomonow, M., Baten, C., Smit, J., Baratta, R., Hermens, H., D'Ambrosia, R., & Shoji, H., (1990). Electromyogram power spectra

frequencies associated with motor unit recruitment strategies. *Journal of Applied Physiology, 68*(3), 1177–1185.

71. Stewart, H. L., & Kawcak, C. E., (2018). The importance of subchondral bone in the pathophysiology of osteoarthritis. *Frontiers in Veterinary Science,* 178.
72. Sturzenegger, M., Beguin, D., Grünig, B., & Jakob, R., (1986). Muscular strength after rupture of the long head of the biceps. *Archives of Orthopedic and Traumatic Surgery, 105*(1), 18–23.
73. Tang, A., & Rymer, W. Z., (1981). Abnormal force--EMG relations in paretic limbs of hemiparetic human subjects. *Journal of Neurology, Neurosurgery & Psychiatry, 44*(8), 690–698.
74. Thomas, S., & Barrington, R., (2003). Hallux valgus. *Current Orthopedics, 17*(4), 299–307.
75. Udall, J. H., Fitzpatrick, M. J., McGarry, M. H., Leba, T. B., & Lee, T. Q., (2009). Effects of flexor-pronator muscle loading on valgus stability of the elbow with an intact, stretched, and resected medial ulnar collateral ligament. *Journal of Shoulder and Elbow Surgery, 18*(5), 773–778.
76. Vanderpool, D., Chalmers, J., Lamb, D., & Whiston, T., (1968). Peripheral compression lesions of the ulnar nerve. *The Journal of Bone and Joint Surgery; British, 50*(4), 792–803.
77. Veeger, H., Kreulen, M., & Smeulders, M., (2004). Mechanical evaluation of the pronator teres rerouting tendon transfer. *Journal of Hand Surgery, 29*(3), 257–262.
78. Veillette, C. J., & Steinmann, S. P., (2008). Olecranon fractures. *Orthopedic Clinics of North America, 39*(2), 229–236.
79. Wadsworth, T. G., (1964). Premature epiphysial fusion after injury of the capitulum. *The Journal of Bone and Joint Surgery; British, 46*(1), 46–49.
80. Warner, J., & McMahon, P. J., (1995). The role of the long head of the biceps brachii in superior stability of the glenohumeral joint. *The Journal of Bone and Joint Surgery; American, 77*(3), 366–372.
81. Williams, M., & Stutzman, L., (1959). *Strength Variation Through the Range of Joint Motion*. In: Oxford University Press.
82. Wilson, N., Ingram, R., Rymaszewski, L., & Miller, J., (1988). Treatment of fractures of the medial epicondyle of the humerus. *Injury, 19*(5), 342–344.

83. Woods, G. W., & Tullos, H. S., (1977). Elbow instability and medial epicondyle fractures. *The American Journal of Sports Medicine, 5*(1), 23–30.
84. Xu, W. D., Gu, Y. D., Liu, J. B., Yu, C., Zhang, C. G., & Xu, J. G., (2005). Pulmonary function after complete unilateral phrenic nerve transection. *Journal of Neurosurgery, 103*(3), 464–467.
85. Yin, D. C. C., (2016). *Functional Outcome of Proximal Fibular Grafting After Wide Resection of distal Radius Tumor in HUSM from Year 2000 to 2013.* Universiti Sains Malaysia.
86. Youm, Y., Dryer, R., Thambyrajah, K., Flatt, A., & Sprague, B., (1979). Biomechanical analyses of forearm pronation-supination and elbow flexion-extension. *Journal of Biomechanics, 12*(4), 245–255.
87. Zhang, Y., & Wang, J., (2018). Classifications of radius and ulna fractures. In: *Clinical Classification in Orthopedics Trauma* (pp. 117–182). Springer.

CHAPTER 4

KINESIOLOGY OF HAND

CONTENTS

4.1. Introduction..106
4.2. Joints and Motions of the Thumb...106
4.3. Joints and Motions of the Fingers 107
4.4. Ligaments and Other Structures 109
4.5. Muscles of the Thumb and Fingers 111
References..117

4.1. INTRODUCTION

The hand is the upper extremity's distal end. It is composed of the metacarpals and phalanges of the thumb and fingers. The hand is the upper extremity's primary point of the function. We utilize our hands to do an endless number of jobs, ranging from the simplest to the most complicated. The upper extremity's other joints' primary function is to position the hand in various postures to complete these activities. The hand is both incredibly helpful and adaptable, as well as being exceedingly complicated. This chapter would just cover the most fundamental function and structure of the hand (Long and Brown, 1964).

4.2. JOINTS AND MOTIONS OF THE THUMB

Thumb joints include metacarpophalangeal (MCP), the carpometacarpal (CMC), and interphalangeal (IP) joints. The trapezium bone expresses with the base of the 1st metacarpal to form the CMC joint. This is a saddle joint with both convex and concave surfaces (Imaeda et al., 1992). The form and connection of such joint surfaces resemble two Pringles potato chips piled on top of each other. The inferior surface of the upper chip resembles the form of the 1st metacarpal, whereas the superior surface of the lower chip resembles the trapezium bone. In one direction, every surface is concave, whereas, in the other, it is convex. The CMC joint is often referred to as a customized ball-and-socket joint, suggesting that it is mobile in all three planes (Figure 4.1) (Cooney et al., 1981).

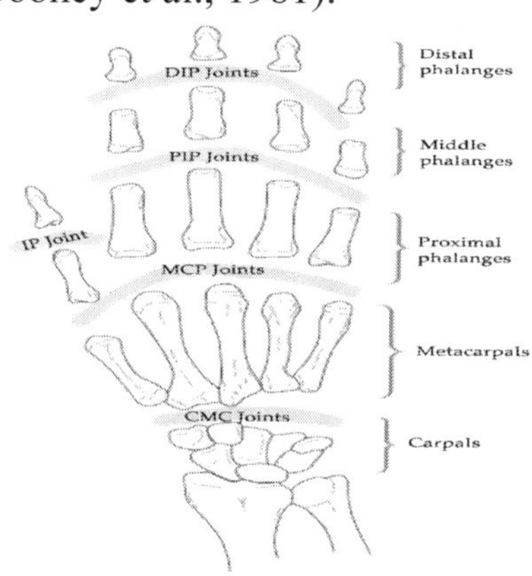

Figure 4.1. Bones and joints of the thumb and fingers.

Source: https://www.researchgate.net/figure/Human-hand-skeletal-structure-depicting-finger-bones-joints-metacarpals-and-carpal_fig2_318184055.

When observing the anatomical posture of your thumb, you would observe that the pad is perpendicular to the palm. Whenever you resist your thumb, the pad now faces the palm or is parallel to it. The rotation has taken place. If you attempt to rotate the thumb without moving any other joints, you would be unable to accomplish it. The rotation at the CMC joint is an involuntary, passive motion caused by the form of the joint. This form of motion is generally known as an auxiliary motion (A motion that follows the active movement and is necessary for normal mobility). The CMC joint of the thumb is more flexible than the CMC joints of the other four fingers, but it offers the same level of stability (Chang and Matsuoka, 2006).

This is uncommon. It permits extension, flexion, adduction, abduction, repositioning, and opposition. Thumb movements are not named in the same manner as other joint motions. Extension and flexion take place in a parallel plane to the palm. Adduction and abduction take place in a plane that is perpendicular to the palm. On the other hand, when the forearm is supinated and the palm is facing up, the side-to-side movement of the thumb across the palm constitutes extension and flexion. The movement of the thumb apart from the palm and up toward the ceiling is abduction, whereas its return is adduction. Repositioning is the return to anatomical position. Opposition is a mix of abduction and flexion with a "built-in" auxiliary motion of rotation. Due to its extra rotation, the CMC joint is typically categorized as a "modified" biaxial joint (Bullock et al., 2012).

The CMC joint of the thumb is very movable, but the IP and MCP joints are not moveable. The MCP joint is a uniaxial hinge joint that permits only extension and flexion. The solitary phalangeal joint, the IP joint, permits only extension and flexion (Li and Tang, 2007).

4.3. JOINTS AND MOTIONS OF THE FINGERS

Each of the 2nd, 3rd, 4th, and 5th digits, usually referred to as the index, middle, and ring fingers have four joints. These have been all the MCP joint, CMC joint, proximal interphalangeal (PIP) and distal interphalangeal (DIP) joints, respectively. CMC joints are nonaxial plane (anomalous) synovial joints that give greater stability than movement (Long and Brown, 1964). As stated, before in the explanation of the thumb joint, the trapezium expresses with the base of the first metacarpal. The trapezoid expresses with the 2nd metacarpal, whereas the capitate and hamate express with the 4th and 5th metacarpals, respectively. The 5th CMC joint of the fingers is the most movable and provides for a limited degree of 5th finger opposition. It does not permit quite so much resistance as the thumb (the 1st CMC joint). The 2nd and 3rd CMC joints are immobile, but the 4th CMC joint is somewhat flexible (Figure 4.2) (Duncan et al., 2013).

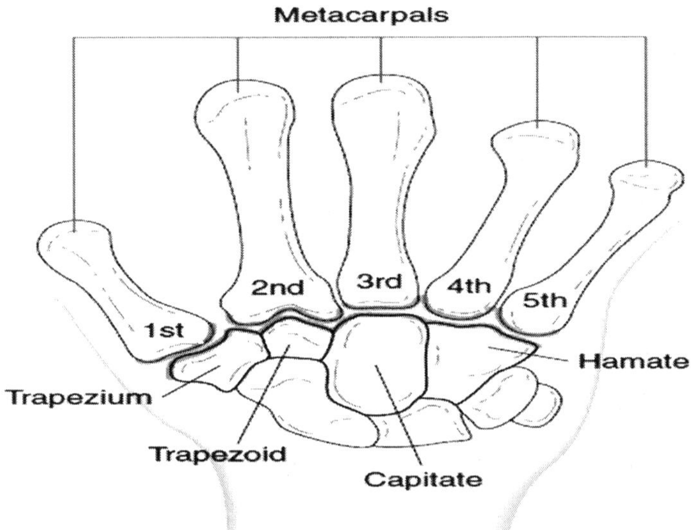

Figure 4.2. The carpometacarpal (CMC) joints of the fingers and thumb.

Source: https://clinicalgate.com/hand-3/.

This may be illustrated by glancing at your knuckles while flexing your elbow and maintaining a supinated position of your forearm. It is important to observe that the hand is in a comfortable palm position, and the MCP joints are almost aligned in a straight line. The 5th MCP joint moves quite a bit and the 4th MCP joint moves to a lesser amount when you create a tight palm, but the 2nd and 3rd MCP joints do not move at all during this motion. The CMC joints are the ones that truly start this MCP movement (Buczek et al., 2011).

The MCP joints, often known as MCP joints, are biaxial condyloid joints that are found in the fingers. The heads of the metacarpals, which are rounded and convex, articulate with the base of the proximal phalanges, which contain a concave form. The term "knuckles" is widely used to refer to certain parts of the body (Nakamura et al., 1998). Such joints can perform the movements of extension, flexion, and hyperextension, in addition to the actions of adduction and abduction. Both adduction and abduction are measured relative to the middle finger as the point of reference. Abduction is the movement of the 2nd, 4th, and 5th fingers away from the middle (3rd) finger, as well as the movement of the middle finger in any direction (Hrabia et al., 2013).

The 2nd, 4th, and 5th fingers are the ones that participate in adduction, which is the return movement from abduction. Only abduction of the middle finger is possible in any direction; there has been no adduction of the middle finger. In each finger, there have been three IP joints. Both the proximal and middle phalanges are connected by the PIP joint, whereas the middle and distal phalanges are connected by the DIP joint. They are hinge joints that only enable extension and flexion and have just one axis of movement (Xu and Todorov, 2016).

4.4. LIGAMENTS AND OTHER STRUCTURES

Even though the hand has countless structures, only a handful of the most well-known would be discussed here. The flexor retinaculum ligament seems to be a fibrous band that runs mediolaterally (horizontally) across the anterior portion of the wrist. Its primary role is to keep such tendons near to the wrist, prohibiting them from bow-stringing (drawing away from the wrist as the wrist bends). It also prohibits the carpal bones' two sides from splitting or expanding apart. This horizontal structure is known as a "tie beam" in the building (Ayhan and Ayhan, 2020). The transverse carpal ligament and the palmar carpal ligament used to be referred to as the transverse and palmar carpal ligaments, respectively. They are recently referred to as the flexor retinaculum. These two elements would be discussed separately due to their clinical importance. The transverse carpal ligament has been more proximal and superficial as compared to the palmar carpal ligament. The transverse carpal ligament blends with its distal fibers. The palmar carpal ligament connects the styloid processes of the ulna and radius to the flexor muscles and crosses them (Moran, 1989; Scapinelli, 1997). The transverse carpal ligament seems to be deeper and distal. It connects to the hamate's pisiform and hooks on the medial side, as well as the trapezium and scaphoid bones on the lateral side. It curves upward over the carpal bones, creating a passageway for the median nerve as well as nine extrinsic flexor tendons that go from the thumb to the fingers (1 tendon for the flexor pollicis longus, 4 tendons for the flexor digitorum profundus, and four tendons for the flexor digitorum superficalis). The fibrous ceiling of the transverse carpal ligament and the bony floor of the carpal bones are both depicted in Figures 4.3. They form the tunnel that the tendons and nerves flow through. The picture also depicts the median nerve's innervation of the hand (Schreuders et al., 2014).

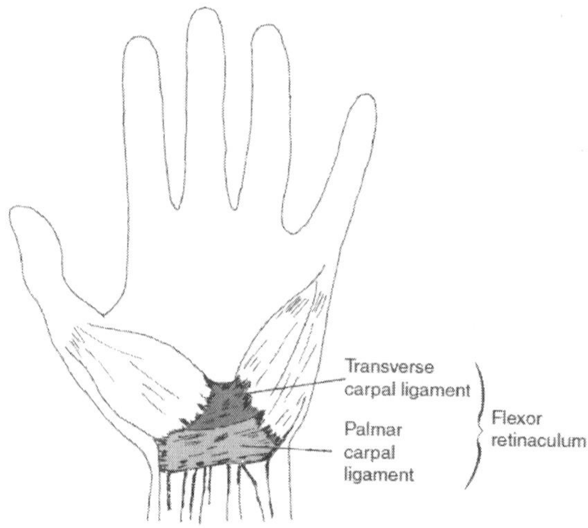

Figure 4.3. Transverse and palmar carpal ligaments constitute the flexor retinaculum.

Source: https://boneandspine.com/flexor-retinaculum/.

The extensor retinaculum ligament is a fibrous band that runs horizontally mediolaterally across the back of the wrist (Subhas et al., 2011). It connects to the ulna's styloid process, as well as the pisiform, triquetrum, and lateral side of the radius, medially. Throughout wrist extension, it keeps the extensor tendons near to the wrist. The extensor hood, also known as the extensor expansion ligament, is a tiny, triangular, flat aponeurosis that covers the dorsum and sides of the proximal phalanx of the fingers (Wells et al., 2013). The tendon of the extensor digitorum mixes with the expansion. It is broader at the base than the MCP joint, and it wraps around the sides. The tendons of the lumbrical and interossei muscles join this as it reaches the PIP joint. At the base of the distal phalanx, it narrows toward its distal end. This expansion connects the lumbrical, extensor digitorum, and interossei muscles to the middle or distal phalanx. Extensor expansion proximally forms an extensor hood that covers the head of the metacarpal and retains the extensor tendon in the midline (Strickland, 2000; Öberg, 1995).

The palm acquires a cupped configuration when it has been relaxed. The arrangement of the skeletal skeleton, which is supported by ligaments, causes the palmar concavity. This structure is made up of three arches. The flexor retinaculum maintains the proximal carpal arch, which is produced by the proximal end of the metacarpals (base) and carpal bones. The metacarpal heads make form the shallower distal carpal arch. For every finger, the longitudinal arch starts at the wrist and runs the length of the metacarpal and phalanges. The remaining two arches are perpendicular to it (Figure 4.4) (Biryukova and Yourovskaya, 1994).

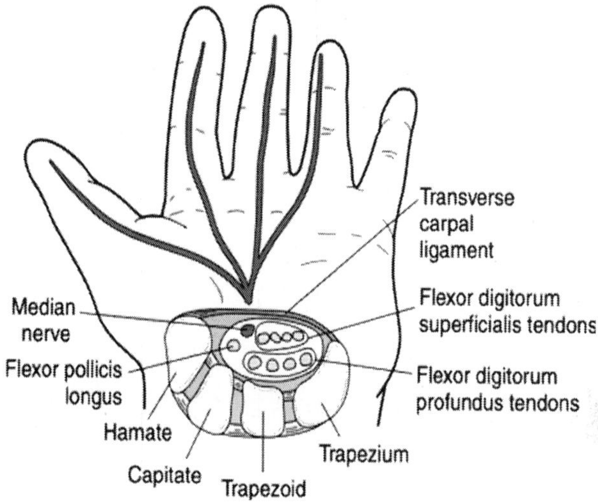

Figure 4.4. The carpal tunnel is composed of the osseous floor of the carpal bones and the fibrous roof of the transverse carpal ligament (anterior superior view). Several tendons including the median nerve run via this tube.

Source: https://www.lecturio.com/concepts/wrist-joint/.

4.5. MUSCLES OF THE THUMB AND FINGERS

4.5.1. Extrinsic Muscles

There have been numerous more muscles that span the wrist and traverse the joints in the hand, in addition to the wrist muscles already discussed. Since their proximal connection is above, or proximal to the wrist joint, such muscles are known as extrinsic muscles of the hand. Their major purpose is at the finger or thumb, although they also help with wrist function. Their names reveal a lot about their purpose and location. Since pollicis means "thumb" in Latin, it is relatively straightforward to recognize the muscles that work on the thumb. The following are the extrinsic muscles (Birdwell et al., 2013):

Anterior	Posterior
Flexor digitorum superficialis	Abductor pollicis longus
Flexor digitorum profundus	Extensor pollicis brevis
Flexor pollicis longus	Extensor pollicis longus
	Extensor digitorum
	Extensor indicis
	Extensor digiti minimi

Deep inside the wrist flexors and palmaris longus muscle is the flexor digitorum superficialis muscle. The typical flexor tendon on the medial epicondyle of the humerus has a large proximal connection. It also contains attachments on the ulna's coronoid procedure and the radius's slant line. It traverses the wrist and separates into four tendons. The distal connection of every finger separates into two segments and connects to every side of the middle phalanx. The PIP and MCP joints of the 2nd through 5th fingers are flexed (Figure 4.5) (Li et al., 2008).

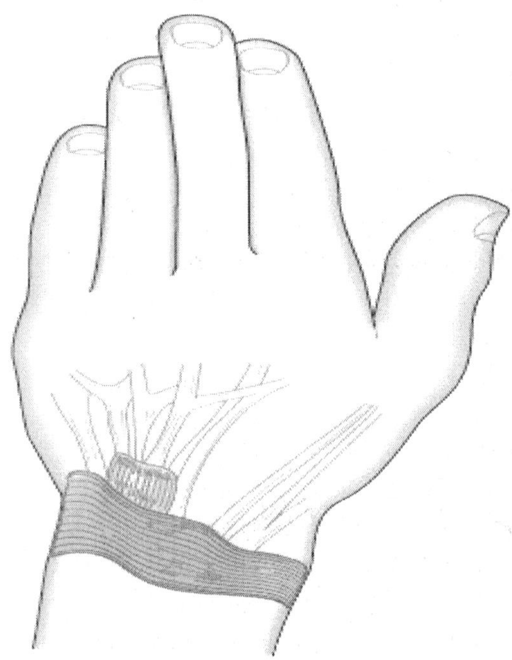

Figure 4.5. Extensor retinaculum (posterior view).

Source: https://www.sciencephoto.com/media/190657/view/the-muscles-of-the-hand.

The flexor digitorum profundus muscle is located deeper than the flexor digitorum superficialis muscle, and the two muscles work jointly to traverse the hand and forearm. From the coronoid procedure to about three-quarters of the way down the ulna, the profundus muscle attaches to the medial and anterior sides of the ulna. It goes beneath the flexor digitorum superficialis muscle till it breaks into two sections at the distal connection of the superficialis tendon (Schieber, 1995). The profundus muscle crosses through this divide and attaches distally at the base of the 2nd through 5th fingers' distal phalanx. The DIP, PIP, and MCP joints of the 2nd through 5th fingers are flexed. The flexor pollicis longus muscle has a proximal attachment on the front side of the radius and the interosseous membrane, and a distal connection at the base of the thumb's distal phalanx. It is a key player in the flexion of the thumb's IP, MCP, and CMC joints (Birdwell et al., 2014).

Deep within the posterior forearm has been the abductor pollicis longus muscle. It joins immediately distal to the interosseous membrane, the supinator, and the middle section of the ulna to the radius. Immediately proximal to crossing the wrist, this becomes superficial and joins to the base of the 1st metacarpal on the radial side. Since the distal joints (IP and MCP) only permit extension and flexion, it essentially abducts the thumb at the CMC joint while being linked solely to the metacarpal (Goehler and Murray, 2010). Consequently, the thumb moves as a unit in the adduction and abduction directions. Adducting the metacarpal likewise adducts the whole thumb. Thus, it is suggested in this text that thumb adduction, abduction, positioning, and opposition happen at the CMC joint when these terms are used (Imaeda et al., 1992).

In addition to its location deep on the extensor pollicis brevis and the posterior forearm muscle spans the wrist just medial to the abductor pollicis longus muscle. It attaches to the posterior radius towards the distal end and slightly below the abductor pollicis longus muscle at its proximal connection. It attaches distally to the posterior surface at the base of the proximal phalanx of the thumb. It acts to expand the thumb's MCP and CMC joints (Vinjamuri et al., 2006). Deeply on the posterior forearm, the extensor pollicis longus muscle is placed near the two described earlier muscles. It is attached to the middle part of the ulna and the interosseous membrane at its proximal end. Before reaching the wrist, this becomes superficial, like the other two muscles. Its distal connection is located at the posterior base of the distal phalanx of the thumb. It works to lengthen the thumb's IP, MCP, and CMC joints (Milner and Dhaliwal, 2002).

4.5.2. Intrinsic Muscles

Intrinsic muscles are those that have their proximal connection at the carpal bones, whereas extrinsic muscles possess their distal attachment at the carpal bones. Intrinsic muscles perform a role on the fingers or thumb. Such muscles are accountable for the fine motor control and precise movement of the hand. It is possible to further subdivide the intrinsic muscles into the deep palm, hypothenar, and thenar muscles. The muscles in the thenar region are the ones responsible for moving the thumb (Royle, 1938). They come together to create the thenar eminence, often known as the ball of the thumb. Between the hypothenar and thenar muscles is where you will find the deep palm muscles. These muscles have been situated deep in the palm. They are responsible for certain of the most complex motions, which often include the use of several different muscles. The adductor pollicis, the interossei (there have been four palmar and four dorsal), and the lumbrical are the names of such muscles (there are four muscles). The little finger is the primary target of the hypothenar muscles, which are responsible for the formation of the hypothenar eminence (Figure 4.6) (Bunnell, 1942).

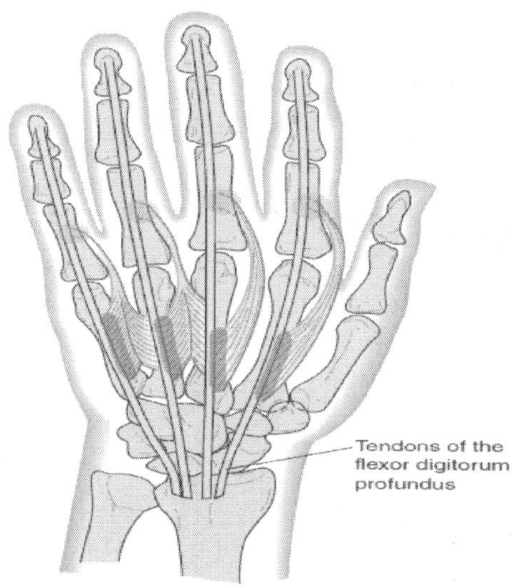

Figure 4.6. The lumbrical muscles (palmar view).

Source: https://www.lecturio.com/concepts/hand/.

The flexor pollicis brevis muscle of the thenar group is rather superficial. It connects proximally to the trapezium and flexor retinaculum and distally to the base of the thumb's proximal phalanx. The major function of this muscle is to flex the MCP and CMC joints of the thumb. Just laterally of the flexor pollicis brevis muscle has been the abductor pollicis Brevis muscle (Brandsma et al., 1995). It connects proximally to the trapezium, scaphoid, flexor retinaculum and distally to the proximal phalanx of the thumb. It functions to abduct the thumb's CMC joint. Deep to the abductor pollicis brevis muscle is the opponens pollicis muscle. It is connected proximally to the trapezium and flexor retinaculum, and distally to the whole lateral surface of the 1st metacarpal (Brandsma et al., 1999).

Its main role is to counteract the thumb. Keep in mind that this activity takes place at the CMC joint. The most crucial function of the hand is thumb resistance. Other muscles, like the abductor pollicis and flexor pollicis brevis, contribute to this function since the thumb combines abduction, rotation, and flexion. The deep palm group, also known as the intermediary group, is made up of muscles positioned between the hypothenar and the thenar muscle groups. Since it is positioned deep into the palm, the adductor pollicis muscle is often included in this group. Due to its effect on the thumb, several sources include it with the thenar group. It is in the deep palm group for no other purpose than to talk about the intrinsic muscles in groups of three (Schreuders et al., 2006)!

Even though it is a thumb muscle, the adductor pollicis is not a commonly regarded portion of the thenar group. It is most likely due to its deep location and lack of muscle mass in the thenar eminence. The capitate, the base of the 2nd metacarpal, and the palmar surface of the 3rd metacarpal are its proximal connections. Its distal connection is located near the base of the thumb's proximal phalanx. Its purpose is to adduct the thumb, as the name indicates (at the CMC joint). The interossei muscles are divided into two groups: palmar and dorsal. The dorsal interossei muscles have been divided into four groups. They connect to two neighboring metacarpals proximally and the base of the proximal phalanx distally (Kilbreath and Gandevia, 1994).

Abducting the 2nd, 3rd, and 4th fingers at the MCP joint is the act that they do. It is important to keep in mind that the 3rd finger might abduct in either direction. The abductor digiti minimi is responsible for the abduction of the 5th finger. The ulnar nerve stimulates the dorsal interossei muscles. There have been four different palmar interossei muscles, just like four different dorsal interossei muscles (Waljee and Chung, 2018). They are attached to the palmar surface of the 1st, 2nd, 4th, and 5th metacarpals, which are located proximally. They are not attached to the middle finger, nor do they serve any purpose on that finger. The distal connection is made to the same finger as the proximal connection, near the base of the proximal phalanx of the similar finger. The ulnar nerve provides the innervation for the palmar interossei muscles, just as it does for the dorsal interossei muscles (Jan and Rooze, 1994).

The middle finger serves as a reference point for adduction and abduction, as previously stated. Abduction is the motion beyond the middle finger, whereas adduction is the movement toward it. Because the middle finger abducts in both directions, it does not adduct. The final muscle group to be addressed is one of a kind. There have been four lubricates, each with its bone attachment. They are fairly deep and only adhere to tendons (Schreuders et al., 2004). They link proximally to the flexor digitorum profundus tendon, which spans the MCP joint anteriorly. They can bend the MCP joint as a result of this. They subsequently travel posteriorly at the proximal phalange, attaching to the extensor digitorum muscle's tendinous expansion. They may now lengthen the DIP and PIP joints. As a result, they flex the MCP joint while extending the DIP and PIP joints of the 2nd to 5th fingers. The "tabletop posture" refers to this combined action. The plural of lumbrical may be spelt with either an "s" or an "es" (Barry et al., 2018).

The hypothenar muscle group is the opposite of the thenar muscle group. With the little finger, the flexor digiti minimi muscle performs a similar function as the flexor pollicis brevis on the thumb. It is connected proximally to the flexor retinaculum and the hamate's hook, and distally to the proximal phalanx of the little finger. It flexes that finger's MCP joint (Johanson et al., 2001). Note that while the CMC joint handles most thumb movement, the MCP joint handles most finger movement. On the ulnar edge of the hypothenar prominence, the abductor digiti minimi muscle sits superficially immediately medial to the flexor digiti minimi muscle. It connects to the pisiform and tendon of the flexor carpi ulnaris muscle proximally, and the base of the proximal phalanx of the 5th finger distally. It abducts that finger's MCP joint (Imaeda et al., 1992).

Deep inside the hypothenar muscles is the opponens digiti minimi muscle. The hook of the hamate and the flexor retinaculum, its proximal connections, are identical to the proximal attachments of the flexor digiti minimi muscle. It connects to the ulnar edge of the 5th metacarpal distally. Its main function is to oppose the 5th finger. This takes place at the CMC joint (Marzke, 1992).

REFERENCES

1. Ayhan, Ç., & Ayhan, E., (2020). Kinesiology of the wrist and the hand. In: *Comparative Kinesiology of the Human Body* (Vol. 1, pp. 211–282). Academic Press.
2. Barry, A. J., Murray, W. M., & Kamper, D. G., (2018). Development of a dynamic index finger and thumb model to study impairment. *Journal of Biomechanics, 77*(1), 206–210.
3. Birdwell, J. A., Hargrove, L. J., & Kuiken, T. A., (2014). Extrinsic finger and thumb muscles command a virtual hand to allow individual finger and grasp control. *IEEE Transactions on Biomedical Engineering, 62*(1), 218–226.
4. Birdwell, J. A., Hargrove, L. J., Kuiken, T. A., & Weir, R. F. F., (2013). Activation of individual extrinsic thumb muscles and compartments of extrinsic finger muscles. *Journal of Neurophysiology, 110*(6), 1385–1392.
5. Biryukova, E. V., & Yourovskaya, V. Z., (1994). A model of human hand dynamics. In: *Advances in the Biomechanics of the Hand and Wrist* (Vol. 1, pp. 107–122). Springer, Boston, MA.
6. Brandsma, J. W., Schreuders, T. A., Birke, J. A., Piefer, A., & Oostendorp, R., (1995). Manual muscle strength testing: Intraobserver and interobserver reliabilities for the intrinsic muscles of the hand. *Journal of Hand Therapy, 8*(3), 185–190.
7. Buczek, F. L., Sinsel, E. W., Gloekler, D. S., Wimer, B. M., Warren, C. M., & Wu, J. Z., (2011). Kinematic performance of a six-degree-of-freedom hand model (6DHand) for use in occupational biomechanics. *Journal of Biomechanics, 44*(9), 1805–1809.
8. Bullock, I. M., Borràs, J., & Dollar, A. M., (2012). Assessing assumptions in kinematic hand models: A review. In: *2012 4th IEEE RAS & EMBS International Conference on Biomedical Robotics and Biomechatronics (BioRob)* (Vol. 1, pp. 139–146). IEEE.
9. Bunnell, S., (1942). Surgery of the intrinsic muscles of the hand other than those producing opposition of the thumb. *JBJS, 24*(1), 1–31.
10. Chang, L. Y., & Matsuoka, Y., (2006). A kinematic thumb model for the ACT hand. In: *Proceedings 2006 IEEE International Conference on Robotics and Automation, 2006: ICRA 2006* (Vol. 1, pp. 1000–1005). IEEE.

11. Cooney, W. P., Lucca, M. J., Chao, E. Y., & Linscheid, R. L., (1981). The Kinesiology of the thumb trapeziometacarpal joint. *J. Bone Joint Surg Am., 63*(9), 1371–1381.
12. Duncan, S. F., Saracevic, C. E., & Kakinoki, R., (2013). Biomechanics of the hand. *Hand Clinics, 29*(4), 483–492.
13. Goehler, C. M., & Murray, W. M., (2010). The sensitivity of endpoint forces produced by the extrinsic muscles of the thumb to posture. *Journal of Biomechanics, 43*(8), 1553–1559.
14. Hrabia, C. E., Wolf, K., & Wilhelm, M., (2013). Whole hand modeling using 8 wearable sensors: Biomechanics for hand pose prediction. In: *Proceedings of the 4th Augmented Human International Conference* (Vol. 1, pp. 21–28).
15. Imaeda, T., An, K. N., & Cooney, III. W. P., (1992). Functional anatomy and biomechanics of the thumb. *Hand Clinics, 8*(1), 9–15.
16. Jan, S. V. S., & Rooze, M., (1994). Anatomical variations of the intrinsic muscles of the thumb. *The Anatomical Record, 238*(1), 131–146.
17. Johanson, M. E., Valero-Cuevas, F. J., & Hentz, V. R., (2001). Activation patterns of the thumb muscles during stable and unstable pinch tasks. *The Journal of Hand Surgery, 26*(4), 698–705.
18. Kilbreath, S. L., & Gandevia, S. C., (1994). Limited independent flexion of the thumb and fingers in human subjects. *The Journal of Physiology, 479*(3), 487–497.
19. Kozin, S. H., Porter, S., Clark, P., & Thoder, J. J., (1999). The contribution of the intrinsic muscles to grip and pinch strength. *The Journal of Hand Surgery, 24*(1), 64–72.
20. Li, Z. M., & Tang, J., (2007). Coordination of thumb joints during opposition. *Journal of Biomechanics, 40*(3), 502–510.
21. Li, Z. M., Tang, J., Chakan, M., & Kaz, R., (2008). Complex, multidimensional thumb movements generated by individual extrinsic muscles. *Journal of Orthopedic Research, 26*(9), 1289–1295.
22. Long, C., & Brown, M. E., (1964). Electromyographic Kinesiology of the hand: Muscles moving the long finger. *JBJS, 46*(8), 1683–1706.
23. Marzke, M. W., (1992). Evolutionary development of the human thumb. *Hand Clinics, 8*(1), 1–8.
24. Milner, T. E., & Dhaliwal, S. S., (2002). Activation of intrinsic and extrinsic finger muscles in relation to the fingertip force vector. *Experimental Brain Research, 146*(2), 197–204.

25. Moran, C. A., (1989). Anatomy of the hand. *Physical Therapy, 69*(12), 1007–1013.

26. Nakamura, M., Miyawaki, C., Matsushita, N., Yagi, R., & Handa, Y., (1998). Analysis of voluntary finger movements during hand tasks by a motion analyzer. *Journal of Electromyography and Kinesiology, 8*(5), 295–303.

27. Öberg, T., (1995). Muscle fatigue and calibration of EMG measurements. *Journal of Electromyography and Kinesiology, 5*(4), 239–243.

28. Royle, N. D., (1938). An operation for paralysis of the intrinsic muscles of the thumb. *Journal of the American Medical Association, 111*(7), 612, 613.

29. Scapinelli, R., (1997). Vascular anatomy of the human cruciate ligaments and surrounding structures. *Clinical Anatomy: The Official Journal of the American Association of Clinical Anatomists and the British Association of Clinical Anatomists, 10*(3), 151–162.

30. Schieber, M. H., (1995). Muscular production of individuated finger movements: The roles of extrinsic finger muscles. *Journal of Neuroscience, 15*(1), 284–297.

31. Schreuders, T. A., Brandsma, J. W., & Stam, H. J., (2014). Functional anatomy and biomechanics of the hand. In: *Hand Function* (Vol. 1, pp. 3–22). Springer, New York, NY.

32. Schreuders, T. A., Roebroeck, M. E., Jaquet, J. B., Hovius, S. E., & Stam, H. J., (2004). Measuring the strength of the intrinsic muscles of the hand in patients with ulnar and median nerve injuries: Reliability of the Rotterdam intrinsic hand myometer (RIHM). *The Journal of Hand Surgery, 29*(2), 318–324.

33. Schreuders, T. A., Selles, R. W., Roebroeck, M. E., & Stam, H. J., (2006). Strength measurements of the intrinsic hand muscles: A review of the development and evaluation of the Rotterdam intrinsic hand myometer. *Journal of Hand Therapy, 19*(4), 393–402.

34. Strickland, J. W., (2000). Development of flexor tendon surgery: Twenty-five years of progress. *Journal of Hand Surgery, 25*(2), 214–235.

35. Subhas, N., Kao, A., Freire, M., Polster, J. M., Obuchowski, N. A., & Winalski, C. S., (2011). MRI of the knee ligaments and menisci: Comparison of isotropic-resolution 3D and conventional 2D fast spin-echo sequences at 3 T. *American Journal of Roentgenology, 197*(2), 442–450.

36. Vinjamuri, R., Mao, Z. H., Sclabassi, R., & Sun, M., (2006). Limitations of surface EMG signals of extrinsic muscles in predicting postures of human hand. In: *2006 International Conference of the IEEE Engineering in Medicine and Biology Society* (Vol. 1, pp. 5491–5494). IEEE.
37. Waljee, J. F., & Chung, K. C., (2018). Surgical management of spasticity of the thumb and fingers. *Hand Clinics, 34*(4), 473–485.
38. Wells, R., Ranney, D., & Keir, P., (2013). Department of Kinesiology university of waterloo. *Advances in the Biomechanics of the Hand and Wrist, 256*(1), 31–35.
39. Xu, Z., & Todorov, E., (2016). Design of a highly biomimetic anthropomorphic robotic hand towards artificial limb regeneration. In: *2016 IEEE International Conference on Robotics and Automation (ICRA)* (Vol. 1, pp. 3485–3492). IEEE.

CHAPTER 5

KINESIOLOGY OF GAIT

CONTENTS

5.1. Introduction 122

5.2. Analysis of Stance Phase 126

5.3. Analysis of Swing Phase 128

5.4. Additional Determinants of Gait 130

5.5. Age-Related Gait Patterns 132

5.6. Abnormal (Atypical) Gait 132

References 141

5.1. INTRODUCTION

Walking is the act of moving from one location to another using your feet. The processor element is called gait. Every person has their distinct style, which can shift significantly depending on their mood. When you are joyful, your steps are lighter and your walk can have a "bounce" to it. Your step can feel heavy when you are depressed or sad, on the other hand. Certain people's walking patterns are so distinctive that they may be recognized from a distance even before their faces are seen. The elements of regular gait remain the same regardless of the countless diverse styles. Walking, at its most primary level, entails balance on one leg as the other travels forward. This necessitates movement of the legs, as well as the trunk and arms. You should first know what joint motions happen to assess gait. You should then determine which muscles or muscle groups are acting depending upon this knowledge (Blanke and Hageman, 1989; Dworak et al., 2010; Kasvand, Milner, Quanbury, and Winter, 1976; Lee, Hyong, and Shim, 2009).

It is necessary to specify the movements of the foot and ankle joint because there is no consensus among authors. Dorsiflexion happens when the dorsal surface of the foot travels towards the anterior surface of the leg. Around the frontal axis, such movements happen in the sagittal plane. Owing to contradictory concepts, extension and flexion must not be employed. Operationally, plantar flexion is identical to the extension because it is a component of the broad extension motion of the knee, ankle, and hip. Thus, technically stated that the plantar flexion is not a real flexion since no two components are brought together. Eversion and inversion refer to motions in the frontal plane about the sagittal axis. Inversion is the process of elevating the medial edge of the foot and inverting the forefoot. In eversion, the lateral edge of the foot is elevated, causing the forefoot to move outward. The movements that occur in the transverse plane are known as adduction and abduction (Verma, Arya, Sharma, and Garg, 2012). Primarily occurring on the forefoot, such movements accompany inversion and eversion, correspondingly (Figure 5.1).

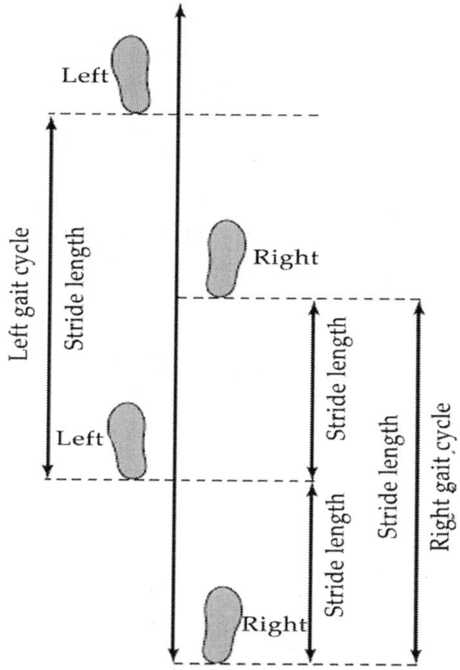

Figure 5.1. Gait cycle terminology. Left or right steps comprise a gait cycle.

Source: https://www.physio-pedia.com/Gait.

Clinicians have started to use the terms pronation and supination to characterize foot and ankle joint movements in current years. Pronation is a composite of eversion, abduction, and dorsiflexion whereas supination is a composite of plantar flexion, adduction, and inversion. Varus and valgus must be specified to avoid additional misunderstanding (Moseley, Smith, Hunt, and Gant, 1996). Such terms are most typically utilized to express a certain posture, generally, one that is aberrant. The position of the distal section distant from the midline is referred to as valgus. Varus, on either hand, is a posture wherein the distal portion is closer to the midline. The distal (inferior) section of the calcaneus is thus tilted away from the middle line, resulting in calcaneal valgus. Such terms would not be utilized here as the focus is on mobility rather than position (Pretterklieber, 1999).

To characterize gait, some definitions should be defined. The gait cycle, also known as stride, is the action that takes place between the time one-foot contacts the ground and the time the similar foot touches the ground again. Throughout During the gait cycle, a stride length is a distance traveled. One-half of a stride is a step. The astride or gait cycle is composed of two steps (one left and one right) (Jordan, Challis, and Newell, 2007; Kharb, Saini, Jain, and Dhiman, 2011; Ren, Jones, and Howard, 2007). These stages must be evenly spaced. A step length is the distance between one foot's heel strike as well as the other foot's heel strike. The step length would decrease or increase as your walking pace increases or decreases. Irrespective of pace, each leg's stride length must be the same. The number of steps done per min is known as cadence. It varies widely. Walking slowly might take up to 70 steps per minute. Students on their way to a test, on either hand, were timed at substantially lower rates. While race walkers would walk significantly faster, fast walking can reach 130 steps in one minute. The gait cycle is similar irrespective of pace; all segments happen in their right position at the appropriate time (Figure 5.2) (Mohammed et al., 2016).

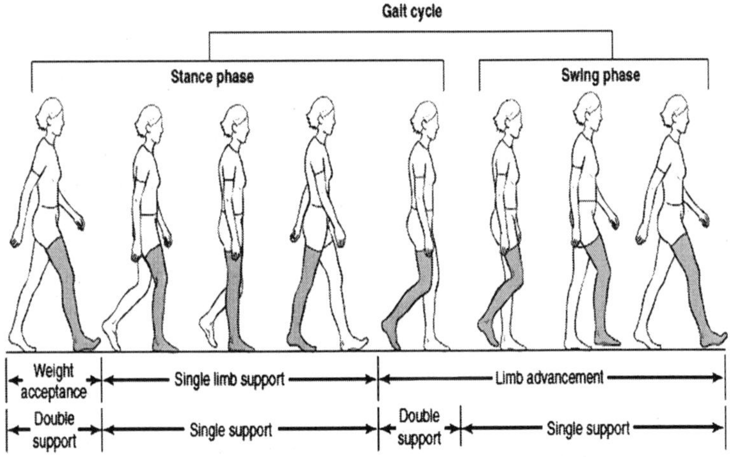

Figure 5.2. The gait cycle's phases.

Source: https://www.orthobullets.com/foot-and-ankle/7001/gait-cycle.

The gait cycle is divided into two stages. The action that happens when the foot makes touches the floor is known as the stance phase. It starts with one foot's heel contact and concludes when that foot leaves the ground. Approximately 60% of the gait cycle occurs during this period (Ivanenko, Poppele, and Lacquaniti, 2004). When the foot is not in contact with the floor, the swing phase begins. It starts when the foot leaves the ground and concludes when the similar foot's heel strikes the ground again. The swing phase accounts for around 40% of the gait cycle (Di Gregorio and Vocenas, 2021).

Throughout these phases of the gait cycle, according to Perry (1992), three goals must be completed: (i) the acceptance of weight; (ii) the support of a single leg; and (iii) the progression of the leg. The stages of the gait cycle are depicted in the diagram. When the foot first contacts the ground and the bodyweight start to move onto that leg, this is when weight acceptance happens (Lai, Leung, Li, and Zhang, 2008). Following then, the entire weight moves totally onto the stance leg, allowing the opposing leg to swing forward. Leg advancement is performed throughout the swing phase.

There are two periods of a single support and two periods of double support in the gait cycle (see Figure 5.2). A phase of double support occurs when both feet are in touch with the ground at the same moment. This happens when one leg starts its stance phase and the other leg finishes its stance phase. The 1st period of double support, for instance, happens when the right leg begins the stance phase and the left leg finishes the stance phase. The 2nd period of double support happens as the right leg's stance phase ends and the left leg's stance phase begins. At an average walking pace, every phase of double support makes up around 10% of the gait cycle. You spend very less time having both feet on the floor whenever you raise your walking pace. Whenever you walk slowly, on either hand, you spend very much time in the double support (Poláková et al., 2020; Wall and Turnbull, 1986; Williams and Martin, 2019).

Walking does not include a nonsupport period or a moment when no foot is in contact with the floor. Nonsupport happens when operating. This is the most significant difference between walking and running, aside from speed. Other actions, like skipping, hopping, and leaping, involve a time of nonsupport but do not follow the same development as running and walking (Nuzir and Dewancker, 2016). To put it another way, such activities do not cover all of the stance and swing phases that running and walking do. Just one foot is in contact with the floor during the single support period (shown in Figure 5.2). In a gait cycle, there are two times of single support: whenever the right foot is on the floor and the left foot swings forward, but when the left foot carries weight and the right leg swings forward. Every support interval accounts for around 40% of the gait cycle (Briskin et al., 2010; Seeber et al., 2014).

To define the elements of walking, several names have evolved from the original, or conventional, vocabulary. Even though the term is true in several circumstances, it is sometimes inconvenient. Furthermore, a language devised by the Rancho Los Amigos (RLA) Medical Center's Gait Laboratory is gaining acceptance. The most significant distinction between the two sets of words is that conventional terms correspond to specific moments in time, while RLA phrases relate to whole eras (Hollman, McDade, and Petersen, 2011; Potluri, Chandran, Diedrich, and Schega, 2019).

RLA periods correctly depict the moving or dynamic aspect of gait, while traditional nomenclature appropriately describes critical moments inside the gait cycle. Since both terminologies would be used in the literature, it is helpful to be acquainted with two sets of words. This section elaborates on what every phase entails and what to look out for.

5.2. ANALYSIS OF STANCE PHASE

As previously stated, the *stance* is the time when the foot is in touch with the ground. Generally, the stance phase consists of five elements:

- Foot flat;
- Mid-stance;
- Heel striking;
- Toe-off; and
- Heel-off.

Several sources list the stance phase as having only four elements, merging heel-off and toe-off into a single element called *push-off*. Since these two periods include very diverse activities, it is essential to maintain them apart.

The start of the stance phase is signaled by a heel strike, which occurs when the heel makes contact with the floor (see Figure 5.3). The ankle is in a neutral posture among plantar flexion and dorsiflexion at this moment, and the knee is starting to flex. Even as the foot strikes the ground, the small knee flexion offers considerable stress absorption. The hip is flexed to around 20° (De Wit, De Clercq, and Aerts, 2000). The trunk is upright and stays that way throughout the gait cycle. The opposing (contralateral) arm is forward, while the same-side (ipsilateral) arm is back in shoulder hyperextension. The body weight starts to move onto the stance leg at this time. This is the phase of the initial encounter in RLA. The ankle dorsiflexors work to keep the ankle in a neutral posture. To reduce the degree of knee flexion, the quadriceps move from concentrically contracting to eccentrically contract (Arnold, Mackintosh, Jones, and Thewlis, 2013).

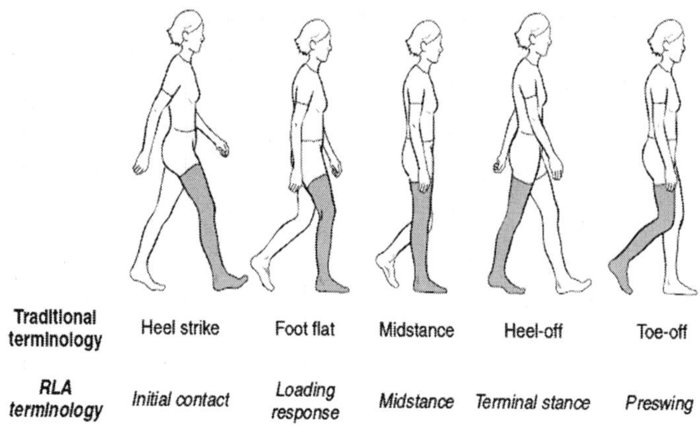

Figure 5.3. The five elements of the stance phase.

Source: https://quizlet.com/197643039/phases-of-gait-cycle-flash-cards/.

Hip flexors are engaged. The hip extensors, on either hand, are starting to contract, preventing the hip from flexing further. The erector spinae work to prevent the trunk from flexing. The force of the foot striking the ground is transmitted to the trunk via the knee, ankle, and hip. If the erector spinae were not there to resist this tension, the pelvis will rotate anteriorly, bending the trunk slightly (Kilbreath, Perkins, Crosbie, and McConnell, 2006; Yu et al., 2008).

When the whole foot hits the floor, the foot flat happens immediately after the heel strike (Figure 5.3). At this point, the ankle is in a neutral position between dorsiflexion and plantar flexion, and the dorsiflexors contract unconventionally to prevent the foot from "slapping" the floor. The knee flexes approximately 20° (Kilbreath et al., 2006; Pirré et al., 2008; Williams, Snow, and Agruss, 1991). The hip is extending, enabling the remainder of the body to start catching up to the leg. Continuation of weight transfer onto the stance leg. Foot flat is analogous to the RLA phase known as loading reaction, which occurs between the conclusion of heel strike and the conclusion of foot flat. Midstance is the point where the body passes over the weight-bearing foot (Figure 5.4). This phase involves modest dorsiflexion of the ankle. The dorsiflexor becomes passive. The plantar flexors contract, regulating the pace of movement of the leg over the ankle. The hip and knee continue expanding; both arms are all in shoulder extension, parallel to the torso, and the trunk is in a balanced rotational posture. Midstance in RLA refers to the time between the midstance end and the foot flat end.

The subsequent midstance is heel-off, where the heel raises off the ground (Figure 5.4). After modest dorsiflexion (about 15°), the ankle would start to plantarflex. That is the start of the push-off stage, also known as the propulsion phase, as the ankle plantar flexors are actively propelling the body forward (Bernardino et al., 2016). The knee is nearly fully extended, the hip is in hyperextension, and the leg is positioned behind the torso. The torso has started to rotate to a similar side, and the arm is moving into shoulder flexion while swinging forward. The final stance in RLA is the interval between the conclusion of midstance and the conclusion of heel-off. Toe-off denotes the conclusion of the push-off section of the stance (Figure 5.4). The MP joints of the toes are in significant hyperextension (Silva et al., 2014). The ankle moves to approximately 10 ° of plantar flexion, while the hip or knee flex. The thigh is vertical to the ground. In RLA, the pre-swing is the time before the toes leave the ground, which signifies the completion of the stance phase and the start of the swing phase.

5.3. ANALYSIS OF SWING PHASE

There are three elements to the swing phase: deceleration, midswing, and acceleration. All of these exercises are non-weight-bearing. The first step is to accelerate. The leg is moving to catch up with the body. The ankle is dorsiflexing, while the hip and knee are still flexing, causing the leg to go forward. In RLA, the first swing is defined as the time between acceleration and toe-off (Figure 5.4) (Anderson, Goldberg, Pandy, and Delp, 2004; Hoy and Zernicke, 1985; Moore et al., 1993; Patil and Chakraborty, 1991).

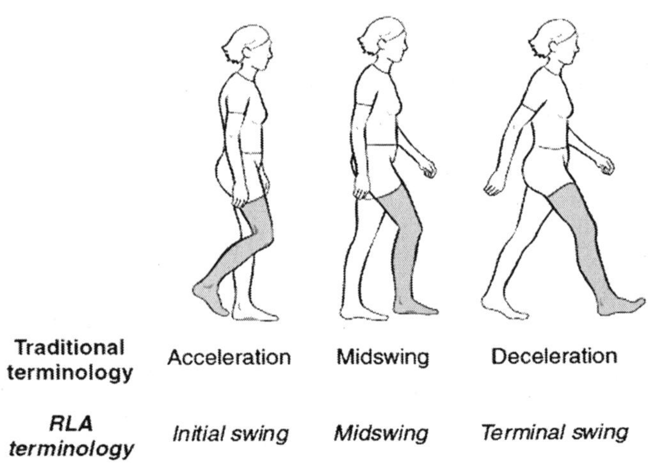

Figure 5.4. Swing phase.

Source: https://musculoskeletalkey.com/assessment-of-gait/.

At the midpoint of the swing, the ankle dorsiflexors have neutralized the ankle. The hip and knee are at their maximal flexion (about 65°). Such movements shorten the leg, enabling the foot to clear the ground during the follow-through swing. Additional hip flexion brings the leg in front of the torso and verticalizes the lower leg.

Midswing is the time between the conclusion of acceleration and the conclusion of midswing in RLA. During deceleration, the ankle dorsiflexors maintain the neutral posture of the ankle in preparation for a heel strike. The hamstring muscles are tightening unconventionally to slow down the leg and prevent it from snapping into extension when the knee extends. The leg has reached its maximum forward position. The remainder inflects. The ultimate swing in RLA is the time between the end of deceleration and the end of the midswing (Figure 5.5) (Alizadeh and Mattes, 2019; Hishikawa et al., 2018).

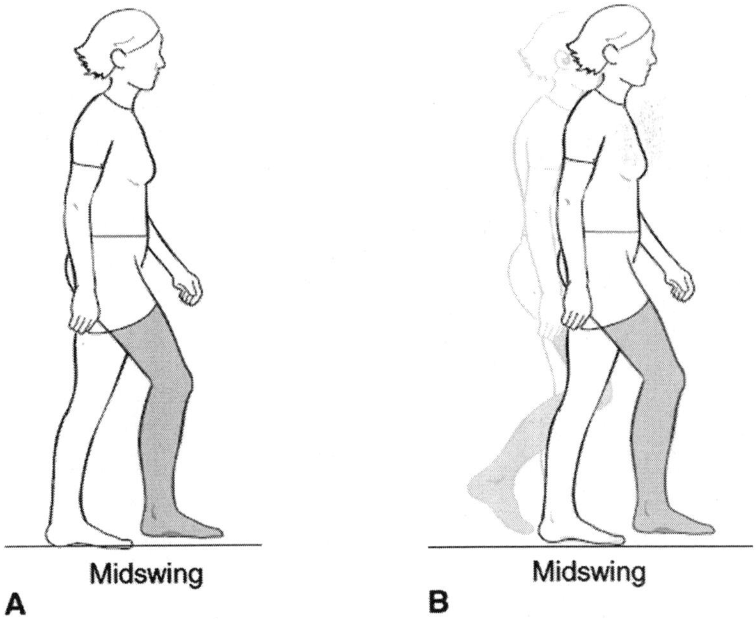

Figure 5.5. (A) Midswing; (B) midswing period (RLA).

Source: https://quizlet.com/197643039/phases-of-gait-cycle-flash-cards/.

5.4. ADDITIONAL DETERMINANTS OF GAIT

Until now, gait analysis has mostly focused on the lower legs. Additional events in the rest of the body, however, should be taken into account. If you traveled the length of the blackboard with a piece of chalk against it, you would notice that the line painted bobs down and up in a wavelike pattern. The vertical displacement of the center of gravity is referred to as this. The usual displacement is around two inches, with the highest point at the midstance and the lowest point at the heel strike (Hishikawa et al., 2018; Kuo, 2007). As the bodyweight varies from side to side, there is an equivalent degree of horizontal displacement of the center of gravity. This displacement is highest at midstance during the single support phase. In the other words, it is the horizontal distance that the body's center of gravity should transfer onto one foot for the other foot to swing forward. The movement from side to side is normally two inches (Figure 5.6) (Della Croce, Riley, Lelas, and Kerrigan, 2001).

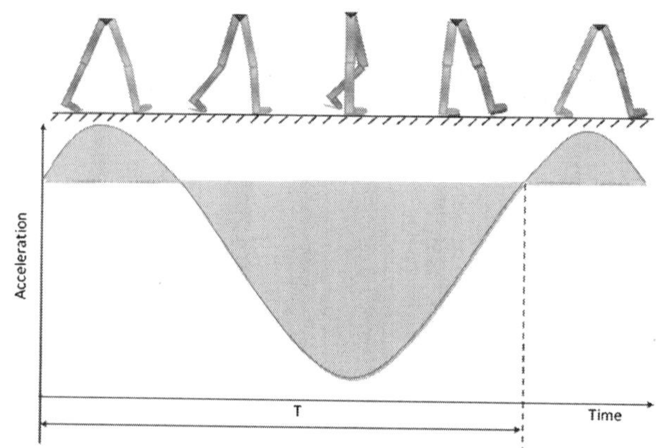

Figure 5.6. Throughout the gait cycle, the body's center of gravity moves vertically.

Source: https://www.hindawi.com/journals/js/2018/4548396/.

When walking, you set your feet significantly apart rather than one in front of the other. This distance will range between two and four inches if lines had been drawn across the sequential midpoints of heel contact (first contact) on every foot. The breadth of the walking base is mentioned below.

If you walked across the room with your hands on your hips, you will observe that they rise and fall as your pelvis lowers down a little on either side. This lateral pelvic tilt, as seen in Figure 5.7, happens when weight is withdrawn from the leg during toe-off (pressing). This modest decline is also known as the Trendelenburg sign (Asayama, Naito, Fujisawa, and Kambe, 2002; Gogu and Gandbhir, 2020). Without the hip abductors on the other side (weight-bearing) and the erector spinae on the similar side working simultaneously, the pelvis will be significantly more slanted. When the right pelvis (the non-weight-bearing side) descends, the left hip (the weight-bearing side) is driven into adduction. Even though the pelvis descends somewhat, the left hip abductors tighten to prevent hip adduction to maintain pelvic symmetry. Simultaneously, the right erector spinal muscle, which is attached to the pelvis, contracts and "pulls up" the side of the pelvis that is also attempting to sink (Ead, Duke, Jaremko, and Westover, 2020; Yildirim et al., 2019).

Furthermore, the length of each step must be comparable in both time and distance. Swing your arms with the opposite leg. As the leg proceeds through the swing phase, the trunk rotates forward. By producing counterrotation, arms swinging in opposite to trunk rotation restrict the amount of trunk rotation. The head must be held high, the shoulders ought to be level, and the trunk must be extended. It is preferable to look at someone's gait from both the side and the front while studying it. Step length, trunk, and head position, arm swing, and lower leg actions are generally best observed from the side. From the front or rear, Examine the width of the walking bass, the posture of the head and shoulders, and the dip of the pelvis.

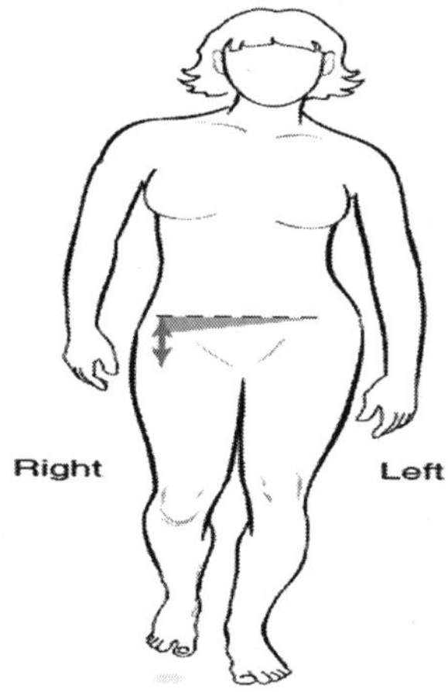

Figure 5.7. Lateral pelvic tilt.

Source: https://www.themedicalmassagelady.co.uk/general-information/massages/sports/injuries/209-lateral-pelvic-tilt.

5.5. AGE-RELATED GAIT PATTERNS

Pathology is not the cause of all gait patterns which do not conform to "normal" gait features. Young children's and adults' walking habits vary markedly from that of younger people. These are age-related variations, not pathological ones (Ead et al., 2020). As children grow older, the distinctions that may be detected in young children begin to disappear. Young children have a broader walking base, a quicker tempo, and a smaller stride length than adults. A flat foot, rather than a heel strike, makes initial contact with the floor (Cupp, Oeffinger, Tylkowski, and Augsburger, 1999). Throughout the stance phase, their knees are generally extended. In other terms, they prefer to take more short, choppy steps in a shorter amount of time. They also have a very limited or non-existent reciprocal arm swing. When a youngster walks alongside an adult, it is obvious to notice (Chester and Wrigley, 2008; Menz, Lord, and Fitzpatrick, 2003; Rowe, Beauchamp, and Wilson, 2021). An aged adult's walking style alters even when there is no disease. Even though there has been no unanimous agreement on the causes of such shifts, fear of falling and security are often believed to be key factors. Generally, elderly people lose muscular mass, become less active, and have worse hearing and eyesight. It is important to remember that the impacts of aging are dependent on a variety of things, including health, exercise level, and even attitude. Certain persons in their 70s may seem "older" than those 10 years more than their senior. Considering all of such caveats, certain broad assertions about the variations in older people's walking patterns may be formed. They take their time walking and spend extra time in the posture phase (Scott, Menz, and Newcombe, 2007).

As a result, the times of double support are longer. Vertical displacement is reduced since they take shorter steps. They walk with a broader base, which results in more horizontal displacement. There have been slower and fewer automatic motions, increasing the likelihood of falling or stumbling. As a consequence of this, there is a potential for enhanced toe-floor clearance.

5.6. ABNORMAL (ATYPICAL) GAIT

An irregular gait can be caused by a variety of factors. It might be temporary, like a sprained ankle, or it could be permanent, as after a stroke. Based on the intensity of the condition, there might be a lot of variety (Colizzi et al., 2020). What is the strength of a muscle? How restricted is joint mobility if it is limited? The intensity or degree of participation, like with other causes of aberrant motion, would always result in a spectrum of small to substantial deviations. There have been several classification systems for abnormal gait (Factor, 2008). The following is a list of aberrant gaits organized by the reason or foundation of the abnormality:

- Restriction of joint or muscle range of motion;
- Muscle paralysis or weakness;
- The disparity in leg length;
- Pain;
- Neurological participation.

5.6.1. Muscular Weakness/Paralysis

Muscle weakness may range from little weakness to total paralysis, in which there has been no strength at all, based on the source or severity of the problem. When muscles weaken, the body compensates by moving the center of gravity over or toward the affected area (Kilic et al., 2018). This lowers the needed muscular strength by reducing the moment of force (torque) on the joint. The part of the gait cycle wherein joints or muscles plays a significant role would be altered. When there has been a discrepancy in terminologies, traditional language would be utilized with RLA phrases in italics (Brusa et al., 2019; Chandar and de Wilton Marsh, 2015; Kilic et al., 2018; Skalova, Minxova, and Slezak, 2008).

At heel strike, the trunk immediately moves posteriorly in the gluteus maximus gait (initial contact). This moves the line of force posterior to the hip joints by shifting the body's center of gravity posteriorly over the gluteus maximus. Because the foot is on the ground, maintaining hip extension during the stance phase demands less muscular strength. Due to the significant backwards-forward motion of the trunk, such shifting is frequently known as a rocking horse gait.

During the stance phase of a gluteus medius gait, the individual rotates their trunk so that it is leaning more toward the afflicted side. The left gluteus medius, also known as the hip abductor, is underdeveloped in Figure 5.8. As a result, the body bends over the left leg during the stance phase of that leg, and the right side of the pelvis descends as the right leg leaves the floor and starts the swing phase. The Trendelenburg gait is another name for this particular manner of walking. It should not be confused with the natural amount of dipping that occurs in the pelvis (Cho, 2019). This maneuver is performed to lessen the amount of strength needed by the gluteus medius to maintain the pelvis by moving the trunk over to the side that is afflicted. Certain potential compensatory strategies can be employed when there has been a weakening in the quadriceps muscle group (Cho, 2019; Lewek, Rudolph, Axe, and Snyder-Mackler, 2002). It is possible to apply a variety of different compensatory maneuvers, but this will depend upon whether or not quadriceps muscles are the only ones that are weak, or whether or not there are further deficits in the extremities. At the beginning of the stance phase, when weight is being transferred onto the stance leg, the individual having quadriceps weakness can lean forward over the quadriceps muscles, which causes the body to lean forward over the quadriceps muscles. In a normal scenario, the line of force will fall behind the knee at this time, necessitating effort from the quadriceps to prevent the knee from giving way. Because bending forward at the hip, the center of gravity will move forward, and the line of force will now be in front of the knee. This will result in the knee being forced into an extended position. To draw the knee into extension upon heel impact, another compensating maneuver involves engaging the hip extensors in conjunction with the plantar flexors in the ankles to create a closed-chain movement. The illustration in figure 5.8 shows how the activity of the muscle may be reversed. During the stance phase, the individual may also apply a physical force on the anterior aspect of the thigh while maintaining a fully extended knee position.

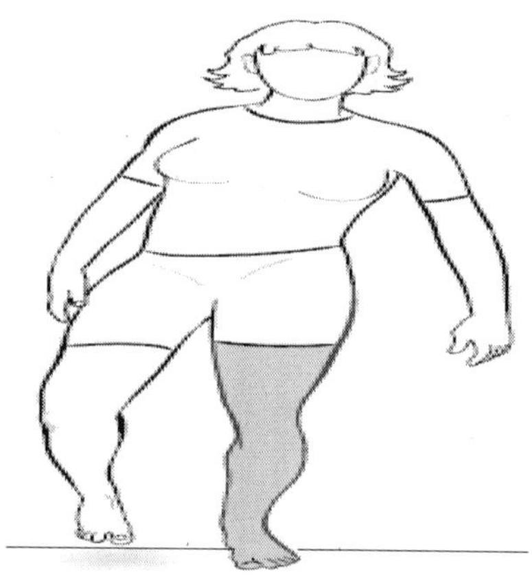

Figure 5.8. Gluteus medius gait.

Source: https://www.corewalking.com/trendelenburg-gait/.

Two things could occur if your hamstrings are weak. The knee would move into extreme hyperextension during the stance phase, which is known as genu recurvatum gait. The knee would snap into extension if the hamstrings do not reduce the forward swing of the lower leg during the deceleration (terminal swing) portion of the swing phase (Fritsch, Dornelles, Oliveira, and Baroni, 2020). The degree of ankle dorsiflexor weakness would impact how a person compensates. The foot would land with a fairly flat foot if there has been inadequate strength to bring the ankle into dorsiflexion at the start of the stance phase. If there has been no ankle dorsiflexion, the toes would hit 1st, resulting in an equinus gait. Following that, weak ankle dorsiflexors can be unable to maintain body weight after heel contact, causing the eccentrical contract to migrate toward the flat foot (loading reaction). A foot slap is an outcome. The foot smashes into plantarflexion when greater weight is placed on the leg because the dorsiflexor is unable to control the drop of the foot. They cannot be capable to dorsiflex the ankle during the swing phase. When the foot is lifted off the ground, gravity causes it to fall into plantarflexion. Drop foot is the term for this (Nicolakis et al., 2000). Consequently, the knee must be raised higher to let the fallen foot clear the floor, resulting in a steppage gait. When playing with a marching band, the drum major would use components of this stride. There is no heel raise during push-off (final stance) whenever the triceps surae group (the soleus and gastrocnemius) is weak, leading to a reduced step length on the unaffected side. A painful foot limp is another name for this condition. This gait is evident even on flat terrain, but it is more noticeable while walking up a hill.

Muscular and other forms of dystrophies frequently create a waddling gait due to the generalized weakening of several muscle groups. The individual stands with the shoulders behind the hips, similar to how a paraplegic will balance on the iliofemoral ligament of the hips (Figure 5.9). Enhanced lumbar lordosis, Trendelenburg gait, and pelvic instability are seen. There is little or no reciprocal rotation of the trunk and pelvis. To forward swing the leg, the complete side of the body should forward swing. Typically, the right arm swings rearward while the right leg swings forward. The right arm and leg swing forward jointly in this instance. This, along with the extreme trunk tilt of Trendelenburg's stride on both sides, reveals the waddling aspect of the gait. A step page gait is frequently observed (Nudds and Codd, 2012; Van Iersel and Mulley, 2004).

Figure 5.9. Gait resulting from quadriceps weakness.

Source: https://www.slideshare.net/HazratBilalmalakandi/human-gait-52124778.

5.6.2. Joint/Muscle Range-of-Motion Limitation

Because of bone fusion or soft tissue constraint, the joint in this classification is unable to move through its usual range of motion. Contractures of the capsule, muscle, or skin might cause this restriction. During the midstance and push-off phases, a person with a hip flexion contracture has been unable to progress into hip hyperextension and extension (end stance) (Soucie et al., 2004). To accommodate, the person would typically take the salutation or greeting stance, wherein the hip has been flexed and the trunk is leaned forward as if bowing. The knee on the concerned limb may also flex when it will ordinarily be extended.

The greater mobility of the lumbar spine and pelvis may considerably accommodate for motion having a fused hip. A lower lordosis and a posterior pelvic tilt enable the leg to swing forward, while a higher lordosis and an anterior pelvic tilt enable the leg to swing backwards. The bell-clapper gait is another name for this. The clapper inside a bell swing back and forth as the bell swings forth and back (Clarkson, 2000; Kisner and Colby, 2012; Rowlett et al., 2019).

A knee flexion contracture would cause excessive dorsiflexion during midstance and a premature heel elevation during push-off (end stance). Additionally, there is a shorter step length on the unaffected side. If a knee fusion is present, the length of the lower leg would be fixed. This length would be based on the joint's location. During the swing phase, the leg would be unable to shorten if the knee is in extension. To accommodate, the individual should (i) raise the toes of the uninvolved leg in a vaulting stride; (ii) lift the hip of the affected side; (iii) swing the leg out to the side; or (iv) use a combination of the aforementioned three techniques. With a circumducted gait, the leg starts near the midline during push-off (final stance), swings outward during the swing phase, and then returns to the midline for heel strike (Ritter et al., 2007). Abducted gait occurs when the leg remains in an abducted posture during the whole walking cycle. Based on the extent of a triceps surae contracture, several possible outcomes might occur. The knee might be driven into extreme extension during midstance if the plantar flexors are too short to permit dorsiflexion. If the gastrocnemius lacks the elasticity to extend over both the knee and ankle, something should give. Either ankle dorsiflexion would be inhibited, or the knee would be forced into excessive extension (Harato et al., 2008). Note that the gastrocnemius is a 2-joint muscle that flexes both the knee and ankle. In weight-bearing, body weight can induce a certain degree of dorsiflexion, so compelling a tight gastrocnemius to extend the knee. Additionally, the heel would rise early during push-off (final stance), the knee would be elevated higher during the swing phase, and the toes would fall before the heel during heel strike (initial contact). This is known as a steppage gait (Jamshidi et al., 2010; Nori, 2019).

Due to the union of the subtalar joint and the two articulations of the transverse joint, ankle fusion is often referred to as a triple arthrodesis. This would lead to a loss of supination and pronation at the ankle. However, plantar flexion and dorsiflexion would be restricted. Typically, a shorter stride length is observed. Losing the capability to supinate and pronate the foot would make it more difficult for the individual to walk on uneven terrain (Figure 5.10) (Easley, Trnka, Schon, and Myerson, 2000; Shah et al., 2017; Ziegler, Friederichs, and Hungerer, 2017).

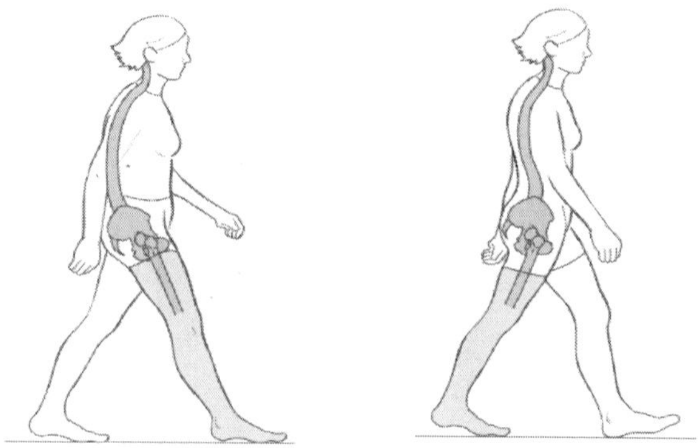

A B

Figure 5.10. (A) The lumbar lordosis is pushed down and the pelvis is tilted posteriorly to achieve the leg swing forward position. (B) By raising lumbar lordosis and tilting the pelvis anteriorly, the leg swings backwards.

Source: https://learnmuscles.com/blog/tag/pelvic-tilt/.

5.6.3. Neurological Involvement

The intensity and extent of neurological involvement determine the level of gait disruption. Spasticity of the hip flexors, for instance, inhibits midstance and final stance leg movement. Hamstring spasticity may cause the knee to remain bent throughout the stance phase, making it difficult to move the leg forward and limiting the efficiency of straightening the leg at the ending of the swing. The triceps surae's spasticity might keep the ankle plantarflexed, causing issues throughout the stance and swing phases. Spasticity causes the foot to be in a varus posture, whereas flaccidity causes the foot to be in a valgus position (Schmid et al., 2013). The intensity of neurological involvement and the existence and quantity of spasticity would determine the hemiplegic gait. Generally, there has been an extension synergy in the affected lower extremities with spasticity. The hip is extended, adducted, and rotated medially. Although the knee is in extension, it is frequently unstable. During both the stance and swing phases, the ankle displays a drop foot with ankle plantarflexion and inversion (equinovarus) (Brunner and Romkes, 2008). Normally, the affected upper extremity is in flexion synergy. There would usually be no reciprocal arm swing. On the involved side, step length seems to be longer, whereas, on the uninvolved side, it seems to be shorter. Cerebellar involvement frequently causes ataxia. Jerky, uneven motions result from a lack of cooperation.

The person who walks has a broader base of support and has a poor balance (abducted gait) (Arnold, Komallu, and Delp, 1997). The individual frequently has trouble walking in a straight line and staggers. The reciprocal arm motion seems choppy and uneven as well. Every movement appears to be exaggerated. A parkinsonian gait is characterized by reduced mobility and tremors. The lower extremities and trunk tend to be flexed. The elbows are slightly flexed, and the reciprocal arm movement is minimal. The forward heel does not swing beyond the back foot, and the stride length is considerably reduced. The person walks with a shuffling stride, with flat feet and most of the weight on the toes. The individual has trouble starting motions. This shuffling stride usually begins slowly and then accelerates, and the person has trouble stopping. It appears as though the person's feet are attempting to keep up with the forward-leaning trunk. A festinating gait is a term for this. A scissors gait is caused by spasms in the hip adductors. During the swing phase, whenever the unsupported leg swings against or across the stance leg, such gait is most noticeable. The walking base has been narrowed. As the swing phase leg tries to swing beyond the stance leg, the trunk can lean over it (Arnold et al., 1997; Uchitomi et al., 2012, 2016).

The bilateral lower extremity impairment found in spastic diplegia linked with cerebral palsy is described as a crouch gait. The gait is frequently different from what is deemed "normal." Hip adduction, flexion, and medial rotation are all excessive, as is knee flexion. Plantar flexion of the ankles. The pelvis retains an anterior pelvic tilt, and lumbar lordosis is enhanced. The reciprocal arm swing and horizontal displacement have been accentuated to compensate (Desrosiers et al., 2003).

5.6.4. Pain

When a person's lower limb joints are in discomfort, they prefer to abbreviate their stance phase. To put it another way, if standing on it hurts, do not do it. On the affected side, a shorter, frequently abducted, stance phase leads to a quick and shortened step length on the uninvolved side. Adjustment is also seen in the reciprocal arm swing. As the step length is reduced, accentuated, and frequently abducted, reciprocal arm swinging shortens. Antalgic gait is a term used to describe this type of walk. If the discomfort is due to a hip condition, the individual would lean over that hip when they bear weight. This reduces the torque on the joint as well as the amount of strain on the femoral head. According to Magee (1987), the amount of pressure applied would be reduced from more than double the person's body weight to nearly the same (Ehde et al., 2000; Gallagher, 2001).

5.6.5. Leg Length Discrepancy

We all have unequally long legs. Frequently, the differences between the left and the right legs might be as great as a quarter inch. How does the body compensate for the discrepancy in leg length, given that both feet must be in touch with the ground in a standing position? In clinical practice, these minor differences have been frequently addressed by placing heel lifts of varying thicknesses into the shoe. In the absence of any further adjustment, lowering the pelvis on the afflicted side (the side with the shorter leg) may accommodate for a slight leg length difference (Shahin et al., 2022). This cannot appear strange, but it places additional strain on the hips, lower back, and knees. In addition to a greater lateral pelvic tilt, the individual can also adjust by leaning over the shorter leg. This will cause the upper body to tilt laterally farther. These procedures may tolerate differences in leg length of up to three inches. When there has been a substantial gap, between three and five inches (based upon the individual's height), lowering the pelvis on the afflicted side is no longer useful. Either the uninvolved leg must be shortened or the involved limb must be made functionally longer. A longer leg is required; hence the affected (shorter) side is often walked on the ball of the foot. This is known as an equine gait. Loss of heel strike (first contact) and foot flat will be the most noticeable change in gait pattern (*loading response*) (Maas, Festen, Hilgenkamp, and Oppewal, 2020; Shahin et al., 2022).

A person may use several strategies to correct for extreme leg length discrepancies (for example, any discrepancy greater than five inches, based on one's height). The person may bend the knee on the unaffected side in addition to lowering the pelvis and walking in an equinus gait. The stance phase will probably start with a flat foot instead of a heel strike in this situation. For the duration of the gait cycle, the knee will be flexed (Defrin, Benyamin, Aldubi, and Pick, 2005). Walk down the street including one leg in the street and the other on the sidewalk to get a sense of how this would feel or appear. It is generally the outcome of some form of the disease when a person's leg length difference exceeds what can be managed using heel lifts. A person with a shattered femur who heals in an overriding posture, for instance, will have a shortened leg. If the epiphyseal plate of one or more long bones in a developing youngster is damaged, that leg's growth may be slowed. If the infant had not yet finished developing, a premature development halt will result in a considerable leg length disparity. These conditions are uncommon, yet they may cause substantial gait abnormalities (Defrin et al., 2005; Vogt, Schiedel, and Rödl, 2014).

REFERENCES

1. Alizadeh, S., & Mattes, K., (2019). How anterior pelvic tilt affects the lower extremity kinematics during the late swing phase in soccer players while running: A time series analysis. *Human Movement Science, 66*, 459–466.
2. Anderson, F. C., Goldberg, S. R., Pandy, M. G., & Delp, S. L., (2004). Contributions of muscle forces and toe-off kinematics to peak knee flexion during the swing phase of normal gait: An induced position analysis. *Journal of Biomechanics, 37*(5), 731–737.
3. Arnold, A. S., Komallu, A. V., & Delp, S. L., (1997). Internal rotation gait: A compensatory mechanism to restore abduction capacity decreased by bone deformity? *Developmental Medicine & Child Neurology, 39*(1), 40–44.
4. Arnold, J. B., Mackintosh, S., Jones, S., & Thewlis, D., (2013). Repeatability of stance phase kinematics from a multi-segment foot model in people aged 50 years and older. *Gait & Posture, 38*(2), 349–351.
5. Asayama, I., Naito, M., Fujisawa, M., & Kambe, T., (2002). Relationship between radiographic measurements of reconstructed hip joint position and the Trendelenburg sign. *The Journal of Arthroplasty, 17*(6), 747–751.
6. Bernardino, G. R., Pelissari, D., Peretti, A. L., Kakihata, C. M. M., Baretta, V., Malanotte, J. A., & Bertolini, G. R. F., (2016). Neural mobilization applied at different sites on the sciatic nerve compression in Wistar rats. *ConScientiae Saúde, 15*(2), 258.
7. Blanke, D. J., & Hageman, P. A., (1989). Comparison of gait of young men and elderly men. *Physical Therapy, 69*(2), 144–148.
8. Briskin, E., Zhoga, V., Chernishev, V., Maloletov, A., Kalinin, Y., & Sharonov, N., (2010). Walking machines (elements of theory, experience of elaboration, application). In: *Emerging Trends in Mobile Robotics* (pp. 769–776). World Scientific.
9. Brunner, R., & Romkes, J., (2008). Abnormal EMG muscle activity during gait in patients without neurological disorders. *Gait & Posture, 27*(3), 399–407.
10. Brusa, C., De Graaf, J., Manzur, A., Main, M., Milev, E., Iodice, M., & Nadarajah, R., (2019). 94 Secondary outcomes of spinal surgery in patients with spinal muscular atrophy (SMA): A retrospective analysis

and a family-centered survey. *Archives of Disease in Childhood, 104*(4), A37.

11. Chandar, M., & De Wilton, M. R., (2015). Severe generalized weakness, paralysis, and aphasia following administration of irinotecan and oxaliplatin during FOLFIRINOX chemotherapy. *Case Reports in Oncology, 8*(1), 138–141.
12. Chester, V. L., & Wrigley, A. T., (2008). The identification of age-related differences in kinetic gait parameters using principal component analysis. *Clinical Biomechanics, 23*(2), 212–220.
13. Cho, M. J., (2019). Lethal cardiac complications in a long-term survivor of spinal muscular atrophy type 1. *Kosin Medical Journal, 34*(1), 47–51.
14. Clarkson, H. M., (2000). *Musculoskeletal Assessment: Joint Range of Motion and Manual Muscle Strength*: Lippincott Williams & Wilkins.
15. Colizzi, M., Ciceri, M. L., Di Gennaro, G., Morari, B., Inglese, A., Gandolfi, M., & Zoccante, L., (2020). Investigating gait, movement, and coordination in children with neurodevelopmental disorders: Is there a role for motor abnormalities in atypical neurodevelopment? *Brain Sciences, 10*(9), 601.
16. Cupp, T., Oeffinger, D., Tylkowski, C., & Augsburger, S., (1999). Age-related kinetic changes in normal pediatrics. *Journal of Pediatric Orthopedics, 19*(4), 475–478.
17. De Wit, B., De Clercq, D., & Aerts, P., (2000). Biomechanical analysis of the stance phase during barefoot and shod running. *Journal of Biomechanics, 33*(3), 269–278.
18. Defrin, R., Benyamin, S. B., Aldubi, R. D., & Pick, C. G., (2005). Conservative correction of leg-length discrepancies of 10mm or less for the relief of chronic low back pain. *Archives of Physical Medicine and Rehabilitation, 86*(11), 2075–2080.
19. Della, C. U., Riley, P. O., Lelas, J. L., & Kerrigan, D. C., (2001). A refined view of the determinants of gait. *Gait & Posture, 14*(2), 79–84.
20. Desrosiers, J., Malouin, F., Richards, C., Bourbonnais, D., Rochette, A., & Bravo, G., (2003). Comparison of changes in upper and lower extremity impairments and disabilities after stroke. *International Journal of Rehabilitation Research, 26*(2), 109–116.
21. Di Gregorio, R., & Vocenas, L., (2021). Identification of gait-cycle phases for prosthesis control. *Biomimetics, 6*(2), 22.

22. Dworak, L. B., Rzepnicka, A., Murawa, M., Maczyński, J., & Buszko, P., (2010). Swing-through gait from the perspective of biomechanics and Kinesiology. Critical analysis of the current state of knowledge and the idea behind the research. *Chirurgia Narzadow Ruchu i Ortopedia Polska, 75*(6), 392–398.
23. Ead, M. S., Duke, K. K., Jaremko, J. L., & Westover, L., (2020). Investigation of pelvic symmetry using CAD software. *Medical & Biological Engineering & Computing, 58*(1), 75–82.
24. Easley, M. E., Trnka, H. J., Schon, L. C., & Myerson, M. S., (2000). Isolated subtalar arthrodesis. *JBJS, 82*(5), 613.
25. Ehde, D. M., Czerniecki, J. M., Smith, D. G., Campbell, K. M., Edwards, W. T., Jensen, M. P., & Robinson, L. R., (2000). Chronic phantom sensations, phantom pain, residual limb pain, and other regional pain after lower limb amputation. *Archives of Physical Medicine and Rehabilitation, 81*(8), 1039–1044.
26. Factor, S. A., (2008). The clinical spectrum of freezing of gait in atypical parkinsonism. *Movement Disorders: Official Journal of the Movement Disorder Society, 23*(S2), S431–S438.
27. Fritsch, C. G., Dornelles, M. P., Oliveira, G. D. S., & Baroni, B. M., (2020). Poor hamstrings-to-quadriceps torque ratios in male soccer players: Weak hamstrings, strong quadriceps, or both? *Sports Biomechanics,* 1–11.
28. Gallagher, D. A., & Malcolm, M. P., (2001). Phantom limb pain and residual limb pain following lower limb amputation: A descriptive analysis. *Disability and Rehabilitation, 23*(12), 522–530.
29. Gogu, S., & Gandbhir, V. N., (2020). *Trendelenburg Sign.*(Vol. 1, pp. 7-12)
30. Harato, K., Nagura, T., Matsumoto, H., Otani, T., Toyama, Y., & Suda, Y., (2008). Knee flexion contracture will lead to mechanical overload in both limbs: A simulation study using gait analysis. *The Knee, 15*(6), 467–472.
31. Hishikawa, N., Tanikawa, H., Ohtsuka, K., Mukaino, M., Inagaki, K., Matsuda, F., & Saitoh, E., (2018). Quantitative assessment of knee extensor thrust, flexed-knee gait, insufficient knee flexion during the swing phase, and medial whip in hemiplegia using three-dimensional treadmill gait analysis. *Topics in Stroke Rehabilitation, 25*(8), 548–553.

32. Hollman, J. H., McDade, E. M., & Petersen, R. C., (2011). Normative spatiotemporal gait parameters in older adults. *Gait & Posture, 34*(1), 111–118.
33. Hoy, M. G., & Zernicke, R. F., (1985). Modulation of limb dynamics in the swing phase of locomotion. *Journal of Biomechanics, 18*(1), 49–60.
34. Ivanenko, Y. P., Poppele, R. E., & Lacquaniti, F., (2004). Five basic muscle activation patterns account for muscle activity during human locomotion. *The Journal of Physiology, 556*(1), 267–282.
35. Jamshidi, N., Rostami, M., Najarian, S., Menhaj, M. B., Saadatnia, M., & Salami, F., (2010). Differences in center of pressure trajectory between normal and steppage gait. *Journal of Research in Medical Sciences: The Official Journal of Isfahan University of Medical Sciences, 15*(1), 33.
36. Jordan, K., Challis, J. H., & Newell, K. M., (2007). Walking speed influences on gait cycle variability. *Gait & Posture, 26*(1), 128–134.
37. Kasvand, T., Milner, M., Quanbury, A., & Winter, D., (1976). Computers and the Kinesiology of gait. *Computers in Biology and Medicine, 6*(2), 111–120.
38. Kharb, A., Saini, V., Jain, Y., & Dhiman, S., (2011). A review of gait cycle and its parameters. *IJCEM International Journal of Computational Engineering & Management, 13*, 78–83.
39. Kilbreath, S. L., Perkins, S., Crosbie, J., & McConnell, J., (2006). Gluteal taping improves hip extension during stance phase of walking following stroke. *Australian Journal of Physiotherapy, 52*(1), 53–56.
40. Kilic, M., Aydin, M. D., Demirci, E., Kilic, B., Yilmaz, I., Tanriverdi, O., & Kanat, A., (2018). Unpublished neuropathologic mechanism behind the muscle weakness/paralysis and gait disturbances induced by sciatic nerve degeneration after spinal subarachnoid hemorrhage: An experimental study. *World Neurosurgery, 119*, e1029–e1034.
41. Kisner, C., & Colby, L., (2012). Range of motion. *Therapeutic Exercise Foundations and Techniques,* 61–73.
42. Kuo, A. D., (2007). The six determinants of gait and the inverted pendulum analogy: A dynamic walking perspective. *Human Movement Science, 26*(4), 617–656.

43. Lai, P. P., Leung, A. K., Li, A. N., & Zhang, M., (2008). Three-dimensional gait analysis of obese adults. *Clinical Biomechanics, 23*, S2–S6.

44. Lee, S. Y., Hyong, I. H., & Shim, J. M., (2009). The effect of aquatic gait training on foot Kinesiology and gait speed in right hemiplegic patients. *The Journal of the Korea Contents Association, 9*(12), 674–682.

45. Lewek, M., Rudolph, K., Axe, M., & Snyder-Mackler, L., (2002). The effect of insufficient quadriceps strength on gait after anterior cruciate ligament reconstruction. *Clinical Biomechanics, 17*(1), 56–63.

46. Maas, S., Festen, D., Hilgenkamp, T. I., & Oppewal, A., (2020). The association between medication use and gait in adults with intellectual disabilities. *Journal of Intellectual Disability Research, 64*(10), 793–803.

47. Menz, H. B., Lord, S. R., & Fitzpatrick, R. C., (2003). Age-related differences in walking stability. *Age and Ageing, 32*(2), 137–142.

48. Mohammed, S., Same, A., Oukhellou, L., Kong, K., Huo, W., & Amirat, Y., (2016). Recognition of gait cycle phases using wearable sensors. *Robotics and Autonomous Systems, 75*, 50–59.

49. Moore, S., Schurr, K., Wales, A., Moseley, A., & Herbert, R., (1993). Observation and analysis of hemiplegic gait: Swing phase. *Australian Journal of Physiotherapy, 39*(4), 271–278.

50. Moseley, L., Smith, R., Hunt, A., & Gant, R., (1996). Three-dimensional kinematics of the rearfoot during the stance phase of walking in normal young adult males. *Clinical Biomechanics, 11*(1), 39–45.

51. Nicolakis, P., Nicolakis, M., Piehslinger, E., Ebenbichler, G., Vachuda, M., Kirtley, C., & Fialka-Moser, V., (2000). Relationship between craniomandibular disorders and poor posture. *CRANIO®, 18*(2), 106–112.

52. Nori, S. L., (2019). *Steppage Gait*.(pp. 1-5)

53. Nudds, R. L., & Codd, J. R., (2012). The metabolic cost of walking on gradients with a waddling gait. *Journal of Experimental Biology, 215*(15), 2579–2585.

54. Nuzir, F. A., & Dewancker, B. J., (2016). Redefining place for walking: A literature review and key-elements conception. *Theoretical and Empirical Researches in Urban Management, 11*(1), 59–76.

55. Patil, K., & Chakraborty, J., (1991). Analysis of a new polycentric above-knee prosthesis with a pneumatic swing phase control. *Journal of Biomechanics, 24*(3, 4), 223–233.
56. Pirré, G. E., Rodini, C., Ferreira, L. T. D., Vasconcelos, J. P. C., & Dos Santos, M. M. C., (2008). Análise da eficiência do treinamento com dinamômetro isocinético no desempenho muscular dos dorsiflexores de um paciente hemiparético espástico, após infiltração de toxina botulínica Tipo A: Estudo de caso. *Acta Fisiátrica, 15*(4), 263–266.
57. Poláková, K., Růžička, E., Jech, R., Kemlink, D., Rusz, J., Miletínová, E., & Brožová, H., (2020). 3D visual cueing shortens the double support phase of the gait cycle in patients with advanced Parkinson's disease treated with DBS of the STN. *Plos One, 15*(12), e0244676.
58. Potluri, S., Chandran, A. B., Diedrich, C., & Schega, L., (2019). Machine learning based human gait segmentation with wearable sensor platform. *Paper Presented at the 2019 41st Annual International Conference of the IEEE Engineering in Medicine and Biology Society (EMBC).*
59. Pretterklieber, M., (1999). Anatomy and kinematics of the human ankle joint. *Der Radiologe, 39*(1), 1–7.
60. Ren, L., Jones, R. K., & Howard, D., (2007). Predictive modelling of human walking over a complete gait cycle. *Journal of Biomechanics, 40*(7), 1567–1574.
61. Ritter, M. A., Lutgring, J. D., Davis, K. E., Berend, M. E., Pierson, J. L., & Meneghini, R. M., (2007). The role of flexion contracture on outcomes in primary total knee arthroplasty. *The Journal of Arthroplasty, 22*(8), 1092–1096.
62. Rowe, E., Beauchamp, M. K., & Wilson, J. A., (2021). Age and sex differences in normative gait patterns. *Gait & Posture, 88*, 109–115.
63. Rowlett, C. A., Hanney, W. J., Pabian, P. S., McArthur, J. H., Rothschild, C. E., & Kolber, M. J., (2019). Efficacy of instrument-assisted soft tissue mobilization in comparison to gastrocnemius-soleus stretching for dorsiflexion range of motion: A randomized controlled trial. *Journal of Bodywork and Movement Therapies, 23*(2), 233–240.
64. Schmid, S., Schweizer, K., Romkes, J., Lorenzetti, S., & Brunner, R., (2013). Secondary gait deviations in patients with and without neurological involvement: A systematic review. *Gait & Posture, 37*(4), 480–493.

65. Scott, G., Menz, H. B., & Newcombe, L., (2007). Age-related differences in foot structure and function. *Gait & Posture, 26*(1), 68–75.
66. Seeber, M., Scherer, R., Wagner, J., Solis-Escalante, T., & Müller-Putz, G. R., (2014). EEG beta suppression and low gamma modulation are different elements of human upright walking. *Frontiers in Human Neuroscience, 8*, 485.
67. Shah, A., Naranje, S., Araoye, I., Elattar, O., Godoy-Santos, A. L., & CESAR, C. D., (2017). Role of bone grafts and bone graft substitutes in isolated subtalar joint arthrodesis. *Acta Ortopédica Brasileira, 25*, 183–187.
68. Shahin, M., Massé, V., Belzile, É., Bédard, L., Angers, M., & Vendittoli, P. A., (2022). Midterm results of titanium conical Wagner stem with challenging femoral anatomy: Survivorship and unique bone remodeling. *Orthopedics & Traumatology: Surgery & Research*, 103242.
69. Silva, D. O., Ferreira, Á. M., Carvalho, A. R., Meireles, A., Tomadon, A., Bertolini, G. R. F., & Marcioli, M. A. R., (2014). Assessment of goniometric acuity of ankle inversion movement: Inter-and intra-raters. *ConScientiae Saúde, 13*(1), 118.
70. Skalova, S., Minxova, L., & Slezak, R., (2008). Hypokalaemic paralysis revealing Sjogren's syndrome in a 16-year old girl. *Ghana Medical Journal, 42*(3), 124.
71. Soucie, J. M., Cianfrini, C., Janco, R. L., Kulkarni, R., Hambleton, J., Evatt, B., & Abshire, T., (2004). Joint range-of-motion limitations among young males with hemophilia: Prevalence and risk factors. *Blood, 103*(7), 2467–2473.
72. Uchitomi, H., Ogawa, K. I., Orimo, S., Wada, Y., & Miyake, Y., (2016). Effect of interpersonal interaction on festinating gait rehabilitation in patients with Parkinson's disease. *Plos One, 11*(6), e0155540.
73. Uchitomi, H., Suzuki, K., Nishi, T., Hove, M. J., Wada, Y., Orimo, S., & Miyake, Y., (2012). Interpersonal synchrony-based dynamic stabilization of the gait rhythm between human and virtual robot—Clinical application to festinating gait of Parkinson's disease patient. *Paper Presented at the 2012 International Symposium on Micro-Nanomechatronics and Human Science (MHS).*

74. Van, I. M. B., & Mulley, G. P., (2004). What is a waddling gait? *Disability and Rehabilitation, 26*(11), 678–682.
75. Verma, R., Arya, K. N., Sharma, P., & Garg, R., (2012). Understanding gait control in post-stroke: Implications for management. *Journal of Bodywork and Movement Therapies, 16*(1), 14–21.
76. Vogt, B., Schiedel, F., & Rödl, R., (2014). Guided growth in children and adolescents. Correction of leg length discrepancies and leg axis deformities. *Der Orthopade, 43*(3), 267–284.
77. Wall, J. C., & Turnbull, G. I., (1986). Gait asymmetries in residual hemiplegia. *Archives of Physical Medicine and Rehabilitation, 67*(8), 550–553.
78. Williams, D. S., & Martin, A. E., (2019). Gait modification when decreasing double support percentage. *Journal of Biomechanics, 92*, 76–83.
79. Williams, K. R., Snow, R., & Agruss, C., (1991). Changes in distance running kinematics with fatigue. *Journal of Applied Biomechanics, 7*(2), 138–162.
80. Yildirim, A., Ayanaoğlu, T., Mustafa, Ö., Esen, E., Kanatli, U., & Bölükbaşi, S., (2019). *The Evaluation of Two Different Surgical Approaches in Total Hip Arthroplasty According to the Patient Satisfaction, Plantar Pressure Distribution and Trendelenburg Sign(Vol.5, pp. 1-10).*
81. Yu, B., Queen, R. M., Abbey, A. N., Liu, Y., Moorman, C. T., & Garrett, W. E., (2008). Hamstring muscle kinematics and activation during overground sprinting. *Journal of Biomechanics, 41*(15), 3121–3126.
82. Ziegler, P., Friederichs, J., & Hungerer, S., (2017). Fusion of the subtalar joint for post-traumatic arthrosis: A study of functional outcomes and non-unions. *International Orthopedics, 41*(7), 1387–1393.

CHAPTER 6

KINESIOLOGY OF NECK AND TRUNK

CONTENTS

6.1. Introduction 150

6.2. Joint Motions 151

6.3. Bones and Landmarks 152

6.4. Joints and Ligaments 156

6.5. Muscles of the Neck and Trunk 158

6.6. Muscles of the Cervical Spine 159

6.7. Muscles of the Trunk 161

6.8. Anatomical Relationships 164

References 166

6.1. INTRODUCTION

The vertebral column creates and manages the body's longitudinal axis. As this is a multijointed pole, the column's movements are caused by the combination of motions of each vertebra. There at cervical area, the spinal column serves as a pivot for mobility and head stabilization. The spinal column transmits the load of the head, shoulder girdle, upper extremities, and trunk. The vertebral column encompasses the spinal cord as well as protects it (Cromwell et al., 2001). Its multijointed pole not just moves, but the placement of such sections also offers good stress absorbing and transfer. The vertebral column is supported by the cranium. The skull is the skeletal framework of the head that contains and protects the brain and facial bones. It is split into bone fragments of the cranium (Kobesova et al., 2014). Since the skull and head contain the sensory systems for vision, listening, tasting, and vestibular reactions, it is critical as the head easily move. This happens as a result of motions at different levels of the cervical spine. The vertebrae are placed in the vertebral column to produce anterior-posterior (AP) (concave-convex) curves which may be viewed from sides. Such curves give the spinal column with around 10 times the strength and determination of a straight rod (Figure 6.1) (Tecco et al., 2011; Öhman, 2015).

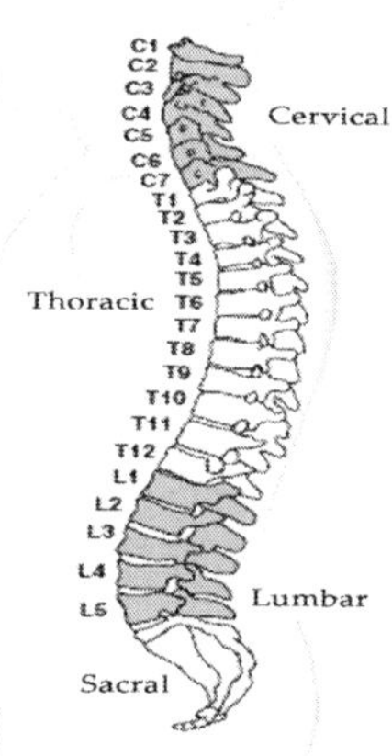

Figure 6.1. The anterior–posterior curves of the vertebral column are seen.

Source: https://www.brainkart.com/article/The-Vertebral-Column---Axial-Skeleton_20916/.

6.2. JOINT MOTIONS

The spinal column is regarded as triaxial overall. As a result, it moves throughout all three dimensions. In the sagittal plane, extension, flexion, as well as hyperextension happen towards the frontal axis (Murray et al., 2016). On the front side, lateral flexing, also known as side bending or lateral flexion, happens itself around the sagittal axis. It still happens on a similar side of the body. Unless from the skull to the atlas, movement happens in the perpendicular direction together around the vertical axis (C1). However, at such a joint, there is no spinning. The number of rotations, as well as other movements available, are largely determined by the position of the facet joints (Figure 6.2) (Bogduk and Yoganandan, 2001).

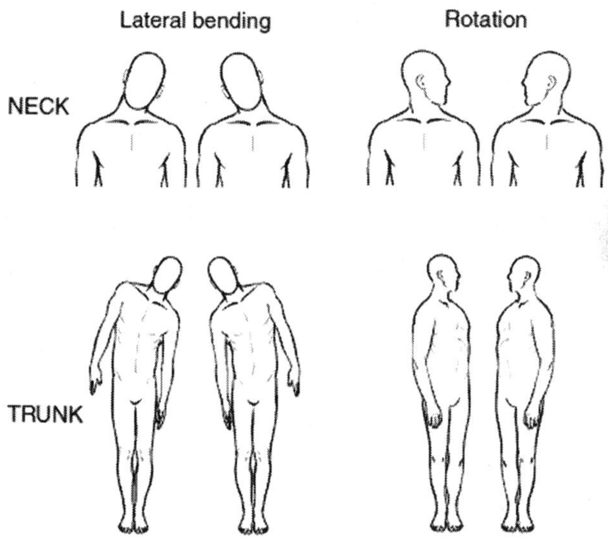

Figure 6.2. Neck and trunk movements.

Source: https://en.hesperian.org/hhg/Disabled_Village_Children:Range-of-Motion_Exercises%E2%80%94Neck_and_Trunk.

The cervical spine provides for head motion and posture and needs more description. The atlantooccipital joint is the articulation from the head to the C1 (atlas). Flexion and extension, as in shaking your head in approval, are the major motions throughout. Among both C1 and C2, there is considerable lateral flexing as well (the atlantoaxial joint) (Emery et al., 2010). The atlantoaxial joint is where most head movement on the neck happens, such as nodding your head in disapproval. The prevertebral muscles superiorly and the suboccipital muscles posterior aspect have the greatest influence over-rotating the head upon that neck. A muscle must get a connection between the skull and the cervical region to mobile the head on the spine (Doriot and Wang, 2006; Ther et al., 2008).

The head (C1) flexes while the neck (C2–C7) extends as you pull your chin in. Axial extension or cervical retraction are terms used to describe this coupled movement. Cervical protraction occurs when the head is extended on C1 and the spine is flexed (C2–C7). Cervical protraction is accentuated by a comfortable front head position or by staring at a computer screen via bifocals. An axial extension is emphasized by "standing up straight" (Woltring et al., 1994).

6.3. BONES AND LANDMARKS

The skull consists of 21 distinct bones and is termed the head's skeleton. Just the bones immediately related to the spinal column would be discussed (Figure 6.3) (Dumas et al., 2007):

- **Occipital Bone:** Also known as the occiput, it constitutes the inferior posterior portion of the skull.
- **Occipital Protuberance:** The tiny protrusion in the middle of the occiput.
- **Nuchal Line:** The horizontal ridge extends with the back of the skull from the occipital protuberance to such mastoid lobes (Huynh et al., 2010).
- **Basilar Area:** The lower or base section of the occiput.
- **Foramen Magnum:** Across a gap in the occipital bone, the spine reaches the skull.
- **Occipital Condyles:** Situated on the occiput laterally to the foramen magnum; facilitates articulation with the atlas (C1) (Thomas et al., 2007).
- **Temporal Bone:** Comprises the foundation and lower lateral edges of the skull.
- **Mastoid Process:** The boney protrusion at the back of the ear whereby the sternocleidomastoid muscle attaches.

Vertebrae (plural of vertebra) vary in size and form but have a similar fundamental arrangement. These are the usual components of a vertebra (Zheng et al., 2012):

- **Body:** Being composed mostly of a cylinder-shaped mass of cancellous bone, the anterior section of the vertebra is the principal load-bearing structure. It is absent from the atlas (C1). From C3 to S1, the size and weight-bearing capacity of the body increase.
- **Neural Arch:** It is the posterior region of the vertebra, also known as the vertebral arch, which is comprised of several distinct elements (Huynh et al., 2015).

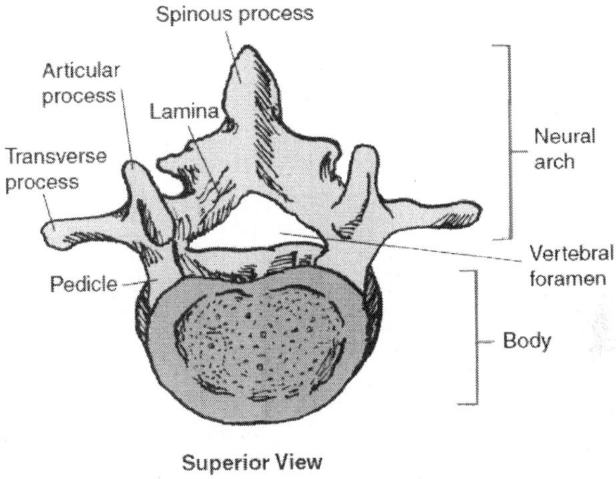

Figure 6.3. The body landmarks of the front and posterior halves of a typical vertebra.

Source: https://basicmedicalkey.com/back/.

- **Vertebral Foramen:** Through the opening generated by the union of the body and neural arch, the spinal cord passes (Huynh et al., 2012).
- **Pedicle:** Right posterior to the body as well as before the lamina, a part of the neural arch.
- **Lamina:** The posterior section of the neural arch joins the two sides at the midline.
- **Transverse Process:** The lateral projections of the arch, which are formed by the junction of the lamina and the pedicle, muscles, and ligaments join together.

- **Vertebral Notches:** These are downturns present on the upper and lower faces of the pedicle (Vasavada et al., 1998).
- **Intervertebral Foramen :** An aperture is created by the upper vertebral groove of the vertebra underneath and the lower vertebral groove of the vertebra above by the notch of the preceding vertebra.
- **Articular Process:** So-called because they project inferiorly and superiorly from the posterior side of every lamina. Superior processes view proximally or longitudinally, while articular elements face posteriorly or laterally (Taylor, 1955).
- **Spinous Process:** Situated at the confluence of the two lamellae, this is the most posterior extension of the neural arch. It acts as a point of attachment for several ligaments and muscles and therefore is palpable along with the entire extent of the vertebral column. An intervertebral disk expresses with neighboring structures across vertebrae. From C2 to C3, 23 discs are starting with the letter C. Their primary purpose is to soak and transfer trauma while preserving the mobility of the vertebral column. About 25% of the total length of the vertebral column is comprised of disks (Wang et al., 2021).
- **Annulus Fibrosus:** The section of the disk's periphery is composed of numerous concentric fibrocartilaginous rings.
- **Nucleus Pulposus:** High-water-content pulpy, gelatinous material in the middle of the disk . By birth, the human body is around 80% water, dropping to below 70% at age 60. This is one reason why individuals lose height as they age. There are a few vertebrae that must be recognized based on their unique properties. They are listed below (Erdmann, 1997):
- **Atlas (C1):** On this first cervical vertebra, the skull sits. As it maintains the head's globe, it is called just after Titan in Greek mythology that supported the world. The atlas is formed like a ring and lacks a body or spinous processes.
- **Anterior Arch:** The front part of the C1 vertebra.
- **Axis (C2):** The second cervical vertebra is so-called because it creates the pivot that enables the head-supporting atlas (C1) to rotate (McBain et al., 2012).

- **Dens:** Also known as the odontoid process; a prominent, anteriorly-located, vertical protrusion of the axis. Into its articulation along with the atlas, cervical movement happens.
- **C7:** Due to its lengthy and conspicuous spinous process, this vertebra is often referred to as the vertebra prominent. It resembles a thoracic vertebra and may be palpated without difficulty when the neck is flexed.
- **Transverse Foramen:** The transverse processes of every cervical vertebra include pores or apertures whereby the vertebral artery travels (Figure 6.4) (George et al., 2017).

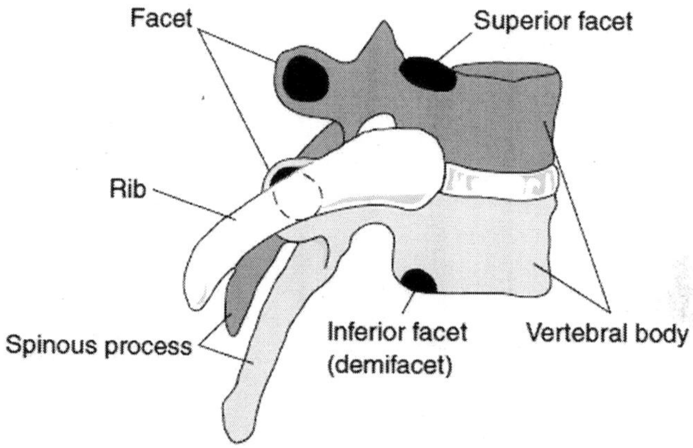

Figure 6.4. The costal facets (rib attachments) of the thoracic vertebrae are seen.

Source: https://healthjade.com/thoracic-vertebrae/.

- **Facet:** They are placed proximally and laterally on the sidewalls of the vertebrae and the lateral aspect of thoracic vertebrae and are also known as coastal facets. The ribs connect to the vertebrae at this point (van der Helm and Veeger, 1996).
- **Demifacet:** In Latin, demi means "half," hence a demifacet is a half or 1/2 face situated horizontally on the upper and lower borders of the vertebral body, in which the ribs articulate with the thoracic vertebrae. Due to the position of the ribs on the body, such borders could include a facet or a demifacet. There are variances between the cervical, thoracic, and lumbar vertebrae despite their shared anatomy (Gayzik et al., 2011).

6.4. JOINTS AND LIGAMENTS

The spinal column starts with two quite dissimilar articulations. The condyles of the occiput engage with the upper articular processes of the axis to produce the atlantooccipital joint. Such connection is sturdy and bears the head's weight (Luo and Goldsmith, 1991). The anterior atlantooccipital barrier is an outgrowth of the fairly narrow posterior longitudinal ligaments. The lateral collateral ligament extends into the tectorial membrane. It supports the spine as it reaches the vertebrae by acting as a sling. The posterior atlantoaxial ligament supports the head's load on the neck. Every one of the condyloid joints created at the confluence of the occipital condyles and the greater trochanter points of the atlas is synovial joints with a synovial membrane surrounded by a joint capsule (Figure 6.5) (Bogduk and Yoganandan, 2001).

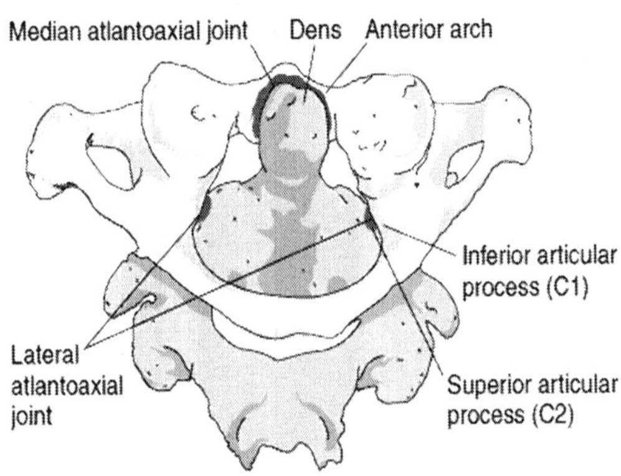

Figure 6.5. Depicts the connection between C1 and C2, illustrating the three atlantoaxial connectors.

Source: https://www.prohealthsys.com/central/anatomy/grays-anatomy/index-5/index-5-3/index-5/atlantoaxial_articulation/.

There are three atlantoaxial points, which are the articulations from the atlas to the axis. The middle atlantoaxial joint is an articulation from the odontoid process (dens) of the axial to the posterior arch of the atlas proximally, and the longitudinal ligaments posteriorly (Happee et al., 2017). Two synovial cavities are observed, one on either end of the dens. Every joint is surrounded by a joint capsule. The anterior and posterior atlantoaxial ligaments are extensions of the posterior and anterior longitudinal ligaments, which span the entirety of the vertebral column. Here between the articular lobes of the two vertebrae are the lateral and medial atlantoaxial junctions (Huynh et al., 2015).

The articulations between C2 and S1 are essentially identical. Powerful, weight-bearing articulations are specific attention between the vertebrae. The posterior region of the vertebra contains two articulations called joint surfaces, on either side (also termed apophyseal or zygapophyseal joints). The face connections are created when the superior articular process of the vertebra beneath articulates with the inferior articular elements of the vertebra beyond (Siegmund et al., 2001).

Every facet joint is a joint with synovium and a joint capsule ligament. Every vertebra is composed of two upper and two lower articular processes. Consequently, every vertebra has two facet joints. Such facet joints significantly define the kind and quantity of mobility available at a particular segment of the vertebral based on the location in which they face. Although lumbar processes are placed in the frontal plane, thoracic processes are situated in the upper extremity lumbar spine experiences the greatest extension and flexion, whereas the thoracic spine experiences the most rotations and lateral flexion. The connection of ribs to vertebrae also adds to the thoracic spine's absence of extension and flexion (Figure 6.6) (Yoganandan et al., 1998).

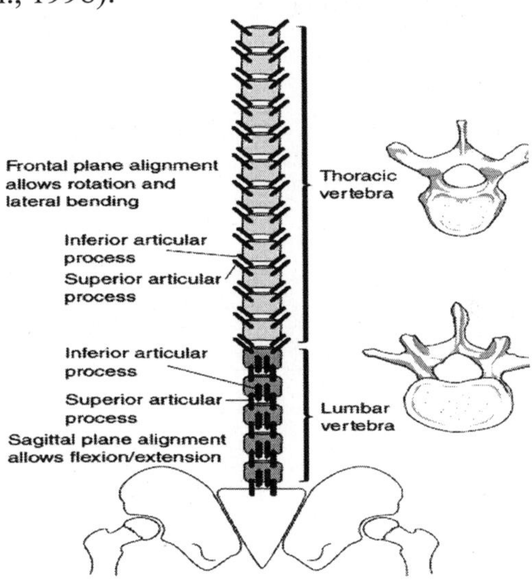

Figure 6.6. The kind of motion is determined by the alignment of the facet joints.

Source: https://www.physio-pedia.com/Thoracic_Anatomy.

As a result of the processes' location from the sagittal to frontal axes on a lateral, the cervical spine has an abundance of all three kinds of mobility.

Numerous ligaments connect all such vertebrae. In the rear area of the body, the posterior longitudinal ligament travels along the vertebral column and tries to avoid severe hyperextension. In which it unites to the sacrum, it is narrow superiorly as well as thick inferiorly (Panjabi et al., 1998). It is located in the thoracic and lumbar areas, immediately behind the aortic arch. The posterior longitudinal ligament runs inferiorly, within the vertebral foramen, with vertebral bodies. The function of this structure is to avoid excessive bending. Proximally, in which it directly supports the head, it is thick. It is narrow inferiorly, contributing to stiffness and enhanced lumbar disk damage. The supraspinal ligament stretches with spinous processes from the 7th cervical vertebra laterally to the sacrum inferiorly (Huynh et al., 2010). The interspinal ligament is situated in adjacent spinous processes. In the cervical area, the extremely strong ligament nuchae (nuchal ligament) replaces the supraspinal and interspinal ligaments. The ligamentum flavum links proximally neighboring laminae (Siegmund et al., 2009).

6.5. MUSCLES OF THE NECK AND TRUNK

The trunk and neck muscles are many and are separated into upper and lower muscles. (There is one limitation: the quadratus lumborum muscle, which is placed in the middle of the plane and is neither a front nor a rear muscle.) The role of the front or rear position has clinical importance. Anterior muscles contract and posterior muscles stretch, just like most other joints. Only the muscles that are medically significant in terms of the exercise would be covered (Figure 6.7) (Cromwell et al., 2001).

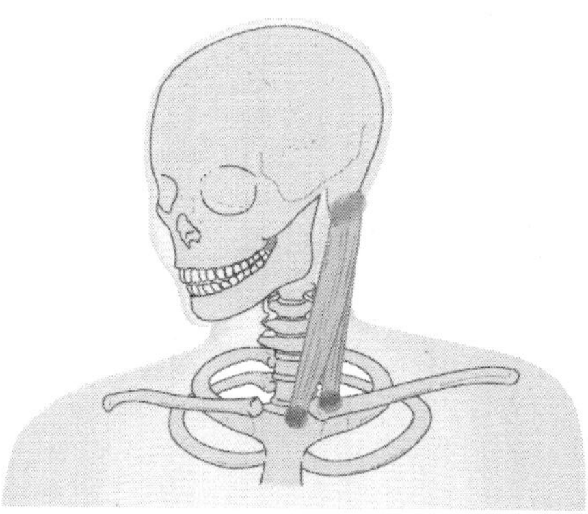

Figure 6.7. The sternocleidomastoid muscle.

Source: https://en.wikipedia.org/wiki/Sternocleidomastoid_muscle.

6.6. MUSCLES OF THE CERVICAL SPINE

In definition, neck flexors are muscles positioned at the front of the cervical spinal column. The sternocleidomastoid muscle, the greatest flexor, is a large, deep, straplike muscle with two divisions originating from the lateral part of the clavicle and the upper end of the sternum. It extends upper and lower to the mastoid process of the temporal bone, where it inserts. Whenever it tightens bilateral, it bends the neck; whenever it bends unilateral, it bends and spins the opposing side of the face horizontally (Bogduk and Yoganandan, 2001). Whenever the right sternocleidomastoid muscle shortens, for instance, your neck turns to the left, causing you to gaze over your left shoulder. Therefore, it turns to the other side. As it is attached to the head, it may influence head movement. Observing the muscle's direction of draw from the side (inferior to the joint axis) reveals as the sternocleidomastoid may hyperextend the head in addition to bending the neck. This emphasizes the "forward head" stance typical of poor posture. Before starting abdominal exercises, one must constantly "tuck the chin" to counteract this motion (Deng and Goldsmith, 1987).

Three scalene muscles are located close to the sternocleidomastoid muscle. The anterior scalene muscle is derived from the lateral aspect of C3 to C6 and enters the anterior part of the first. The mid scalene muscle is derived from the transverse processes of C2 to C7 and enters into the superior part of the first rib (Ivancic, 2012). The posterior scalene muscle, the shortest and most superficial muscle, arises from the cervical vertebrae C5 to C7 and enters into the second rib. Due to their proximity and the fact that they all serve similar function, it is unnecessary to discriminate amongst them. They are positioned laterally in the neck and are particularly efficient at flexing the cervical spine laterally. Due to their proximity to the plane, they only aid in flexion (Figure 6.8) (Huelke and Nusholtz, 1986).

There is a collection of anterior muscles often known as the prevertebral muscles. They are deeply situated and extend with the anterior cervical vertebrae. Such muscles are responsible for neck and head flexion. Due to their diminutive size in comparison to other neck flexors, their most important function may be preserving postural stability and "tucking" the chin. Multiple tiny neck muscles act as supports for the hyoid bone and tongue (Emery et al., 2010).

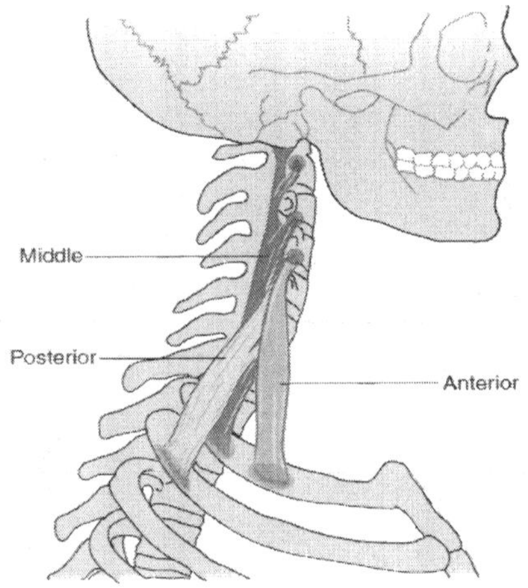

Figure 6.8. Depicts the three scalene muscle components.

Source: https://www.kenhub.com/en/library/anatomy/scalene-muscles.

The hyoid bone is distinctive in that it lacks bony joints. It serves as the major stability for the tongue's many muscles. Such muscles have little effect on the movements of the cervical spine. From across all dimensions, such muscles reach the bottom of the skull (Lavallee et al., 2013).

The suboccipital muscles are grouped posteriorly behind the base of the skull and solely lift the skull. The muscles cooperate to expand the head by moving the occipital condyles on the atlas or to turn it by rotating the skull and atlas and skull around the odontoid process of the axes (Nightingale et al., 2016).

The basic muscles with the posterior spinal column are classified as the erector spinae family, which will be explored in more depth alongside the trunk muscles in the following subsections. Such muscles give postural control over the gravitational force that causes the head to flex; also serve as extensors to return the head to its original posture. The lowest spinal muscles (transversospinalis, interspinal, and intertransversarii) would be discussed in the part on the trunk since the bulk of such muscles is situated there (Barrett et al., 2020).

The splenius capitis and splenius cervicis muscles are located close to the erector spinae. As indicated by their titles, they connect to the head and cervical spine. Splenius capitis is the most basic of the two muscles. They adhere to the posterior aspect of the lower cervical and upper thoracic vertebrae and extend proximally and horizontally to the lateral occiput (capitis) and the longitudinal elements of the higher cervical vertebrae (cervicis), respectively (Luo and Goldsmith, 1991). Whenever muscles on just one side tighten, the neck and the face rotate and fold horizontally to a similar side. Since either ends flex, therefore, they stretch the neck, while the splenius capitis stretches the head on the neck. In some circumstances, the upper trapezius and levator scapula may aid the splenius capitis and cervicis. If the scapula is stabilized, the muscles may work in the other direction. Rather than the neck and head rotating on the scapula, the neck and head rotate on the scapula (Gosselin et al., 2004).

6.7. MUSCLES OF THE TRUNK

The rectus abdominis muscle runs across the anterior trunk at the midline. The linea alba separates the two sections from one another. The rectus abdominis muscle comes from the pubic crest and enters into the fifth, sixth, and seventh ribs' costal cartilages. The muscle is divided laterally into separate components by three tendon crossings. The rectus abdominis muscle, which is positioned in the anterior centerline, is a powerful trunk bender that, along with other anterior trunk muscles, compresses the abdominal organs (Emery et al., 2010).

Observe how the trunk flexes on the hips while executing a sit-up. If the legs or ankle are pulled down, the hip flexor muscle is also implicated in a reversal of muscular activity while completing a sit-up. The knees as well as hips ought to be bent, but the ankles must not be kept down if the goal is to develop the abdominals (not the hip flexors). Hip flexors are shortened when the knees and hips are flexed, giving them less effective. Whenever the distal portion (feet or legs) is not maintained, the hip flexors may not reverse muscular activity (held down) (Kier and Smith, 1985).

The external oblique muscle is a wide, broad, flattened muscle on the anterolateral abdomen that lays externally. It starts on the bottom eight ribs and continues posterior aspect and obliquely to the iliac crest and, through the abdominal aponeurosis, to the linea alba in the middle. The fibers of the right and left external oblique muscles to create a V when they are combined. Once both ends flex, the trunk flexes and the abdominal contents are compressed. Whenever one end tightens, the external oblique flexes horizontally to the opposing side as well as spins the trunk (Kier and Smith, 1985).

The right external oblique muscle turns the right-hand side of the trunk to the center in this position. Consider shifting your right shoulder forwards and to the left. The oblique muscles are found beneath and perpendicularly to the outer oblique muscle. The inguinal ligament, iliac crest, and thoracolumbar fascia all contribute to that too. So, it passes proximally and laterally into the three consecutive ribcages and through the linea alba through the abdominal aponeurosis (Tecco et al., 2011). The fibers of the right or left internal oblique muscles combine to produce an inverted V shape. Whenever either side engages, they bend and squeeze the abdominal organs, similar to the outer oblique muscle. Whenever one side flexes, the trunk folds to the side of the inner oblique. The internal oblique muscle, on the other hand, has the inverse result of rotating, turning the trunk to a similar side (Nashner, 2014). The right obliques muscle turns the right-hand side of the trunk away from the middle in that position. Consider shifting your right shoulder back as well as to the right. As a result, when it comes to twisting the trunk to the left, the right outer slant and left inner oblique are antagonists. The left exterior and right interior obliques are antagonistic over the similar motion (Mork and Westgaard, 2009).

The transverse abdominis muscle, which sits underneath the inner oblique muscle, is the most superficial abdominal muscle. It is called for the parallel or longitudinal orientation of its fibers. It is derived from the lateral aspect of the inguinal ligament, the iliac crest, the thoracolumbar fascia, and the final six bones. It directly traverses the belly to enter into the abdominal aponeurosis and linea alba. Due to its straight axis of pull, it contributes nothing to the movement of the trunk. Nevertheless, it collaborates with other abdominal muscles to squeeze and maintain the abdominal organs (Kobesova et al., 2014). This is crucial for actions like sneezing, coughing, laughter, forceful exhalation, and "bearing down" in birthing or bowel movements. There are several types of back muscles. Certain generalizations may be established about their connections and behaviors. For instance, muscles connecting spinous processes to spinous processes have a vertical line of pull; hence, they lengthen. Due to their central location, there is just one pair of these. The column of pull for muscles that go from the anterior aspect to the transverse process is medial to the midline. They flex horizontally while operating singly and stretch while operating bilaterally (Vasseljen and Westgaard, 1997).

The line of pull for muscles connecting from rib to rib is identical to that of muscles connecting across transverse processes. Become more horizontal, rib-attaching muscles are more efficient in lateral flexion than other muscles. Muscles connecting from the spinous process to the transverse process or vice versa have had an angled line as well as consequently stretch symmetrically and spin unilaterally. Smaller muscles are much more efficient at spin, whereas larger muscles are more successful in elongation (Swartz et al., 2005). Sometimes referred to as the sacrospinalis muscle group, the erector spinae muscles comprise the interlayer of back extensor muscles. That muscle collection may be split into three groups running along the spinal column and linking the spinous processes, transverse processes, as well as ribs (Park and Srinivasan, 2021).

One of the most medial muscle groups is the spinal, which connects predominantly to the nuchal ligament and spinous processes of the cervical and thoracic vertebrae. That group's component that connects to the occiput also connects to the lateral aspect of the cervical vertebrae. Those muscles, positioned in the middle, are the primary actuators in trunk extension. The longissimus muscle group is placed laterally to the spinal group of muscles and attaches to the transverse processes extending from the occiput to the sacrum (St-Onge et al., 2011). Since such muscles are medial to the centerline as well as have a diagonal line of force, they squeeze unilaterally to induce medial bending and symmetrically to produce extension. The iliocostalis muscles are the most lateral muscle group, typically connecting posteriorly to the ribs. They adhere proximally to transverse processes and laterally to the sacrum and ilium. Due to their location on the side of the body, those muscles are efficient in lateral flexion. Bilaterally functioning, they are efficient extensors (Perchthaler et al., 2015). Those three muscular groups are often referred to as the erector spinae muscle group; consequently, they will be discussed together. The higher fibers of the spinalis and longissimus groups connect to the occiput and may consequently extend the head on the neck from a transverse process to the spinous process of a vertebra beyond; hence, they are very efficient at spinning. The semispinalis muscles often cover five or more vertebrae, the multifidus muscles typically span two to four vertebrae, and the rotatores muscles, which are the smallest and most superficial of this group, span just one vertebra (O'Sullivan et al., 2006).

These muscles lengthen the spine and twist to the opposing side. That collective's most basic muscle is the semispinalis. The rotators are the lowest of such muscles, with the multifidus lying beneath it. Those two muscles, like the transversospinalis muscle group, are situated deeply, but their line of pull is vertical rather than oblique. Thus, they should be evaluated independently. The titles of the interspinal and intertransversarii muscles reveal their attachment points (Ekstrom et al., 2007). In the majority of the vertebral column, the interspinal muscles connect from the spinous process here to the spinous process beyond. With just this vertical pull line in the middle, these muscles are excellent extensors. The intertransversarii muscles are located along with the majority of the vertebral column and connect the transverse processes behind and in front of. They are efficient in flexing laterally. The quadratus lumborum muscle arises from the iliac crest and is a major muscle (Gosselin et al., 2004). It inserts superiorly onto the last rib and transverse elements of all lumbar vertebrae. Since it is placed in the AP centerline, it has no bending or expansion function, and since it is straight, it has no rotational function. Nevertheless, having medial to the midline enables it to flex laterally. It serves another purpose whenever its base is pushed to its placement (reversal of muscle action). This is referred to as hip hiking or the lifting with one part of the pelvis. It is a crucial feature for anybody with a long leg cast or a fused knee since it enables the feet to clean the ground with no need of bending the knee (Bogduk and Yoganandan, 2001).

6.8. ANATOMICAL RELATIONSHIPS

The platysma muscle, which is extremely wide and narrow, is the most basic in the occipital region. A substantial amount of the two lateral necks is covered by this muscle. Which does not have a function at the neck and contributes to facial expression. The sternocleidomastoid muscle is located under the platysma and runs crosswise from the lateral clavicle out now and upward to the mastoid process at the back of the ear (Swartz et al., 2005). The infrahyoid muscles are located deep in the sternocleidomastoid and align more vertical in the anterior neck area. The suprahyoid muscles may be seen under the chin. The prevertebral muscles, which lie adjacent to the spinal column, are the deepest muscle group (not visible). The platysma covers all except the top half of the sternocleidomastoid when seen from the side. The sternocleidomastoid covers the infrahyoid muscles anteriorly and the three scalene muscles along the midline laterally because it travels diagonally from posterior-superior to anterior-inferior (Burdi et al., 1969).

The posterior scalene is hidden from view. Near their superior attachments, the sternocleidomastoid covers sections of the levator scapula and splenius capitis. The posteriorly most superficial muscle is the upper trapezius. The posterior neck has many layers of muscles. The upper trapezius, as previously stated, is the most superficial muscle. The figure 6.8 shows the splenius capitis partly covering the splenius cervicus by removing this muscle. The semispinalis muscle of the transversospinalis group lies underneath these muscles. The erector spinae are included in this category (not visible). The shortest muscles are found in the deepest region of the neck.

The interspinales and intertransversarii muscles, as well as the suboccipital (near the head) muscles. These latter two muscles are hidden (Zhang et al., 2006).

The anterior and posterior muscles of the trunk are separated. The abdominal, or anterior, trunk wall has four layers of muscles. The superficial Rectus femoris is located in the midline. The external oblique muscle is located directly under the rectus femoris anteriorly, on the sidewalls of the abdominal wall. The internal oblique is located just underneath the external oblique muscle. The transverse abdominis is the deepest abdominal muscle, with fibers that travel horizontally. Deep inside the shoulder girdle and shoulder joint muscles are the posterior trunk muscles (Deng and Goldsmith, 1987). The erector spinae muscles, which include the iliocostalis (lateral column), longissimus (middle column), and spinalis, are the most superficial layer of back muscles, as shown in figure 6.8 (medial column). The intrinsic back muscles of the transversospinalis group are located deep inside the erector spinae muscles. These muscles run vertically between the transverse and spinous processes in the groove. The one-joint interspinal and intertransversarii muscles are the deepest in the trunk (Tecco et al., 2011).

REFERENCES

1. Barrett, J. M., McKinnon, C., & Callaghan, J. P., (2020). Cervical spine joint loading with neck flexion. *Ergonomics, 63*(1), 101–108.
2. Bogduk, N., & Yoganandan, N., (2001). Biomechanics of the cervical spine part 3: Minor injuries. *Clinical Biomechanics, 16*(4), 267–275.
3. Burdi, A. R., Huelke, D. F., Snyder, R. G., & Lowrey, G. H., (1969). Infants and children in the adult world of automobile safety design: Pediatric and anatomical considerations for design of child restraints. *Journal of Biomechanics, 2*(3), 267–280.
4. Cromwell, R. L., Aadland-Monahan, T. K., Nelson, A. T., Stern-Sylvestre, S. M., & Seder, B., (2001). Sagittal plane analysis of head, neck, and trunk kinematics and electromyographic activity during locomotion. *Journal of Orthopedic & Sports Physical Therapy, 31*(5), 255–262.
5. Deng, Y. C., & Goldsmith, W., (1987). Response of a human head/neck/upper-torso replica to dynamic loading—II. Analytical/numerical model. *Journal of Biomechanics, 20*(5), 487–497.
6. Doriot, N., & Wang, X., (2006). Effects of age and gender on maximum voluntary range of motion of the upper body joints. *Ergonomics, 49*(3), 269–281.
7. Dumas, R., Cheze, L., & Verriest, J. P., (2007). Adjustments to McConville et al. and Young et al. body segment inertial parameters. *Journal of Biomechanics, 40*(3), 543–553.
8. Ekstrom, R. A., Donatelli, R. A., & Carp, K. C., (2007). Electromyographic analysis of core trunk, hip, and thigh muscles during 9 rehabilitation exercises. *Journal of Orthopedic & Sports Physical Therapy, 37*(12), 754–762.
9. Emery, K., De Serres, S. J., McMillan, A., & Côté, J. N., (2010). The effects of a Pilates training program on arm–trunk posture and movement. *Clinical Biomechanics, 25*(2), 124–130.
10. Erdmann, W. S., (1997). Geometric and inertial data of the trunk in adult males. *Journal of Biomechanics, 30*(7), 679–688.
11. Gayzik, F. S., Moreno, D. P., Geer, C. P., Wuertzer, S. D., Martin, R. S., & Stitzel, J. D., (2011). Development of a full body CAD dataset for computational modeling: A multi-modality approach. *Annals of Biomedical Engineering, 39*(10), 2568–2583.

12. George, N. C., Kahelin, C., Burkhart, T. A., & Andrews, D. M., (2017). Reliability of head, neck, and trunk anthropometric measurements used for predicting segment tissue masses in living humans. *Journal of Applied Biomechanics, 33*(5), 373–378.
13. Gosselin, G., Rassoulian, H., & Brown, I., (2004). Effects of neck extensor muscles fatigue on balance. *Clinical Biomechanics, 19*(5), 473–479.
14. Happee, R., De Bruijn, E., Forbes, P. A., & Van, D. H. F. C., (2017). Dynamic head-neck stabilization and modulation with perturbation bandwidth investigated using a multisegment neuromuscular model. *Journal of Biomechanics, 58*(1), 203–211.
15. Huelke, D. F., & Nusholtz, G. S., (1986). Cervical spine biomechanics: A review of the literature. *Journal of Orthopedic Research, 4*(2), 232–245.
16. Huynh, K. T., Gibson, I., & Gao, Z., (2012). Development of a detailed human spine model with haptic interface. *Haptics Rendering and Applications, 1*, 165–193.
17. Huynh, K. T., Gibson, I., Jagdish, B. N., & Lu, W. F., (2015). Development and validation of a discretized multi-body spine model in LifeMOD for biodynamic behavior simulation. *Computer Methods in Biomechanics and Biomedical Engineering, 18*(2), 175–184.
18. Huynh, K. T., Gibson, I., Lu, W. F., & Jagdish, B. N., (2010). Simulating dynamics of thoracolumbar spine derived from LifeMOD under haptic forces. *World Academy of Science, Engineering and Technology, 4*(4), 236–243.
19. Ivancic, P. C., (2012). Biomechanics of sports-induced axial-compression injuries of the neck. *Journal of Athletic Training, 47*(5), 489–497.
20. Kier, W. M., & Smith, K. K., (1985). Tongues, tentacles and trunks: The biomechanics of movement in muscular-hydrostats. *Zoological Journal of the Linnean Society, 83*(4), 307–324.
21. Kobesova, A., Valouchova, P., & Kolar, P., (2014). Dynamic neuromuscular stabilization: Exercises based on developmental Kinesiology models. *Functional Training Handbook* (Vol. 1, pp. 25–51).
22. Lavallee, A. V., Ching, R. P., & Nuckley, D. J., (2013). Developmental biomechanics of neck musculature. *Journal of Biomechanics, 46*(3), 527–534.

23. Luo, Z., & Goldsmith, W., (1991). Reaction of a human head/neck/ torso system to shock. *Journal of Biomechanics, 24*(7), 499–510.
24. McBain, K., Shrier, I., Shultz, R., Meeuwisse, W. H., Klügl, M., Garza, D., & Matheson, G. O., (2012). Prevention of sports injury I: A systematic review of applied biomechanics and physiology outcomes research. *British Journal of Sports Medicine, 46*(3), 169–173.
25. Mork, P. J., & Westgaard, R. H., (2009). Back posture and low back muscle activity in female computer workers: A field study. *Clinical Biomechanics, 24*(2), 169–175.
26. Murray, M., Lange, B., Chreiteh, S. S., Olsen, H. B., Nørnberg, B. R., Boyle, E., & Sjøgaard, G., (2016). Neck and shoulder muscle activity and posture among helicopter pilots and crew-members during military helicopter flight. *Journal of Electromyography and Kinesiology, 27*, 10–17.
27. Nashner, L. M., (2014). Practical biomechanics and physiology of balance. *Balance Function Assessment and Management, 1*, 431.
28. Nightingale, R. W., Sganga, J., Cutcliffe, H., & Cameron, R., (2016). Impact responses of the cervical spine: A computational study of the effects of muscle activity, torso constraint, and pre-flexion. *Journal of Biomechanics, 49*(4), 558–564.
29. O'Sullivan, P., Dankaerts, W., Burnett, A., Straker, L., Bargon, G., Moloney, N., & Tsang, S., (2006). Lumbopelvic kinematics and trunk muscle activity during sitting on stable and unstable surfaces. *Journal of Orthopedic & Sports Physical Therapy, 36*(1), 19–25.
30. Öhman, A., (2015). The immediate effect of Kinesiology taping on muscular imbalance in the lateral flexors of the neck in infants: A randomized masked study. *PM&R, 7*(5), 494–498.
31. Panjabi, M. M., Cholewicki, J., Nibu, K., Babat, L. B., & Dvorak, J., (1998). Simulation of whiplash trauma using whole cervical spine specimens. *Spine, 23*(1), 17–24.
32. Park, J. H., & Srinivasan, D., (2021). The effects of prolonged sitting, standing, and an alternating sit-stand pattern on trunk mechanical stiffness, trunk muscle activation and low back discomfort. *Ergonomics, 64*(8), 983–994.
33. Perchthaler, D., Hauser, S., Heitkamp, H. C., Hein, T., & Grau, S., (2015). Acute effects of whole-body vibration on trunk and neck

muscle activity in consideration of different vibration loads. *Journal of Sports Science & Medicine, 14*(1), 155.

34. Siegmund, G. P., Myers, B. S., Davis, M. B., Bohnet, H. F., & Winkelstein, B. A., (2001). Mechanical evidence of cervical facet capsule injury during whiplash: A cadaveric study using combined shear, compression, and extension loading. *Spine, 26*(19), 2095–2101.
35. Siegmund, G. P., Winkelstein, B. A., Ivancic, P. C., Svensson, M. Y., & Vasavada, A., (2009). The anatomy and biomechanics of acute and chronic whiplash injury. *Traffic Injury Prevention, 10*(2), 101–112.
36. St-Onge, N., Côté, J. N., Preuss, R. A., Patenaude, I., & Fung, J., (2011). Direction-dependent neck and trunk postural reactions during sitting. *Journal of Electromyography and Kinesiology, 21*(6), 904–912.
37. Swartz, E. E., Floyd, R. T., & Cendoma, M., (2005). Cervical spine functional anatomy and the biomechanics of injury due to compressive loading. *Journal of Athletic Training, 40*(3), 155.
38. Taylor, C. L., (1955). The biomechanics of control in upper-extremity prostheses. *Artif. Limbs, 2*(3), 4–25.
39. Tecco, S., Crincoli, V., Di Bisceglie, B., Caputi, S., & Festa, F., (2011). Relation between facial morphology on lateral skull radiographs and sEMG activity of head, neck, and trunk muscles in Caucasian adult females. *Journal of Electromyography and Kinesiology, 21*(2), 298–310.
40. Tecco, S., Mummolo, S., Marchetti, E., Tetè, S., Campanella, V., Gatto, R., & Marzo, G., (2011). sEMG activity of masticatory, neck, and trunk muscles during the treatment of scoliosis with functional braces. A longitudinal controlled study. *Journal of Electromyography and Kinesiology, 21*(6), 885–892.
41. Ther, J. M. P., (2008). Applied Kinesiology research and literature compendium: Neck pain caused by muscle weakness. *J. Manipulative Physiol. Ther., 31*(7), 518–524.
42. Thomas, J. S., France, C. R., Sha, D., Vander, W. N., Moenter, S., & Swank, K., (2007). Applied Kinesiology research and literature compendium: Low back pain caused by muscle weakness. *Spine, 32*(26), E801–808.
43. Van, D. H. F. C., &Veeger, H. E. J., (1996). Quasi-static analysis of muscle forces in the shoulder mechanism during wheelchair propulsion. *Journal of Biomechanics, 29*(1), 39–52.

44. Vasavada, A. N., Li, S., & Delp, S. L., (1998). Influence of muscle morphometry and moment arms on the moment-generating capacity of human neck muscles. *Spine, 23*(4), 412–422.

45. Vasseljen, Jr. O., & Westgaard, R. H., (1997). Arm and trunk posture during work in relation to shoulder and neck pain and trapezius activity. *Clinical Biomechanics, 12*(1), 22–31.

46. Wang, J., Siddicky, S. F., Carroll, J. L., Rabenhorst, B. M., Bumpass, D. B., Whitaker, B. N., & Mannen, E. M., (2021). Infant inclined sleep product safety: A model for using biomechanics to explore safe infant product design. *Journal of biomechanics, 128*(1), 110706–110710.

47. Woltring, H. J., Long, K., Osterbauer, P. J., & Fuhr, A. W., (1994). Instantaneous helical axis estimation from 3-D video data in neck kinematics for whiplash diagnostics. *Journal of biomechanics, 27*(12), 1415–1432.

48. Yoganandan, N., Pintar, F. A., Cusick, J. F., & Kleinberger, M., (1998). Head-neck biomechanics in simulated rear impact. In: *Annual Proceedings/Association for the Advancement of Automotive Medicine* (Vol. 42, p. 209). Association for the Advancement of Automotive Medicine.

49. Zhang, Q. H., Teo, E. C., Ng, H. W., & Lee, V. S., (2006). Finite element analysis of moment-rotation relationships for human cervical spine. *Journal of Biomechanics, 39*(1), 189–193.

50. Zheng, L., Jahn, J., & Vasavada, A. N., (2012). Sagittal plane kinematics of the adult hyoid bone. *Journal of Biomechanics, 45*(3), 531–536.

CHAPTER 7

A KINESIOLOGY OF HIP JOINT

CONTENTS

7.1. Introduction 172
7.2. Joint Motions and Structure 172
7.3. Landmarks and Bones 174
7.4. Ligaments and Other Structures 177
7.5. Muscles of the Hip 181
7.6. Anatomical Relationships 184
7.7. Common Hip Pathologies 185
7.8. An Overview of Muscle Innervation 187
References 188

7.1. INTRODUCTION

The foot, leg, thigh, and pelvis make up the lower extremity. The two bones of the hip (os coxae bones), the coccyx and the sacrum are the pelvis bones. The bone of the hip is made up of three fused bones (pubis, ischium, and ilium). The patella and femur are found in the thigh. The fibula and tibia are part of the leg, whereas the foot has five metatarsals, 7 tarsal bones, and 14 phalanges (Figure 7.1) (Floyd and Thompson, 2009; Neumann, 2010).

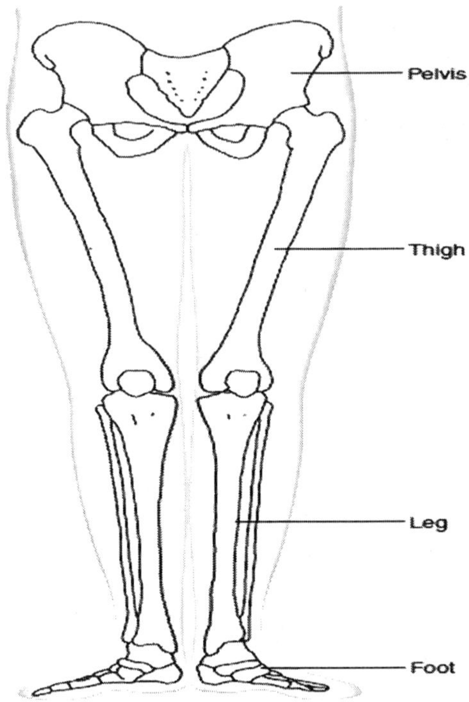

Figure 7.1. The bones of the lower extremities.

Source: https://www.wesnorman.com/llbones.htm.

7.2. JOINT MOTIONS AND STRUCTURE

The hip has been the most proximal joint of the lower extremities. It is crucial for weight-bearing and walking activities. Similar to the shoulder, it consists of a ball-and-socket joint. The rounded or convex head of the femur fits into and expresses with the concave acetabulum. The convex femoral head glides in the reverse direction of the thigh's motion.

In contrast to the shoulder, the hip is a fairly solid joint that compromises considerable mobility. In contrast, the shoulder is less stable since it is highly mobile (Rutherford et al., 2015; Eyüboğlu et al., 2020). As a triaxial joint, the hip is capable of movement in all three planes. In the flexion, sagittal plane, hyperextension, and extension happen, comprising roughly 15° and 120° of hyperextension. Extinction is the opposite of flexion. In the frontal plane, abduction and adduction occur, with approximately 45° of abduction. Typically, adduction is considered the opposite of anatomical position, however, an extra 25° of motion is conceivable beyond the anatomical position. Lateral and medial rotations in the transverse plane are commonly known as exterior and interior rotations, correspondingly (Figure 7.2) (Dostal et al., 1986; Harput, 2020).

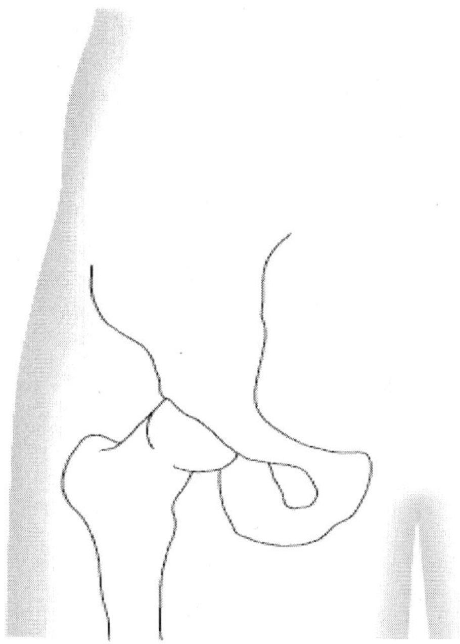

Figure 7.2. The hip joint (anterior view).

Source: https://www.physio-pedia.com/Hip_Anatomy.

From the anatomical position, roughly 45° of rotation is available in every direction. The two bones of the hip are joined anteriorly to each other and posteriorly to the sacrum. Additionally, the sacrum is attached distally to the coccyx. The pelvic pelvis girdle is comprised of such four bones (the two hip bones, the coccyx, and the sacrum). Remember that the femur has not included in the pelvis (Kaneda et al., 2012; Rallis et al., 2018; Morcelli et al., 2016).

7.3. LANDMARKS AND BONES

As previously stated, the hip joint consists of the femur and the hip bone. The hip bone, commonly referred to as the os coxae, is comprised of three unevenly shaped bones: the ischium, the pubis, and the ileum. By maturity, these bones have fused. The superior part of the hip bone comprises the fan-shaped ilium (Figures 7.3 and 7.4) (Murray et al., 1975; Rosner and Cuthbert, 2012). Important landmarks include the following:

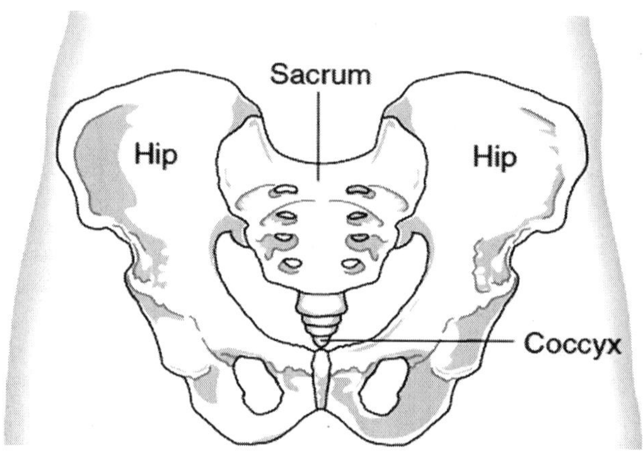

Figure 7.3. The bones of the pelvis (anterior view).

Source: https://www.researchgate.net/figure/Anterior-view-of-the-pelvic-girdle-Adapted-from-Van-de-Graaff-2001-Ch-7_fig1_216453509.

1. **Iliac Fossa:** The iliac component of the iliopsoas muscle joins to this smooth, large, concave region on the interior surface (Harty, 1984).
2. **Iliac Crest:** As you place your hands on your hips, your hands rest on this bony portion. The posterior superior iliac spine (PSIS) and the anterior superior iliac spine (ASIS) define their limits.
3. **Anterior Superior Iliac Spine (ASIS):** There is a protrusion at the anterior extremity of the iliac crest. Here are the attachments of the sartorius, the tensor fascia latae, and the inguinal ligament (Nonaka et al., 2002; Lee et al., 2014).

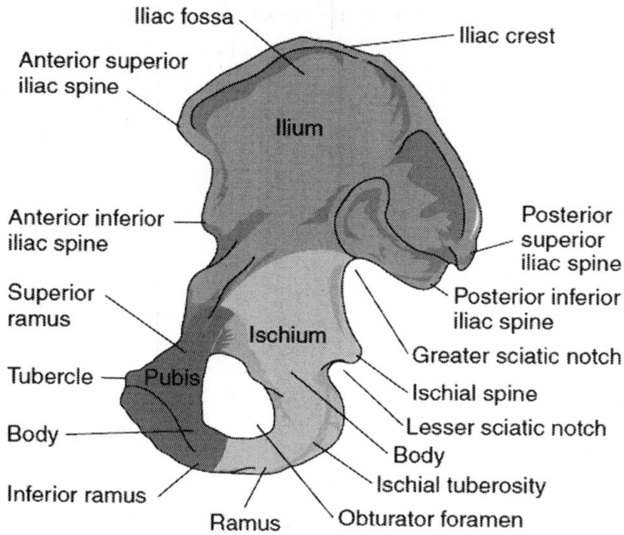

Figure 7.4. Right hip bone.

Source: https://teachmeanatomy.info/pelvis/bones/hip-bone/.

4. **Anterior Inferior Iliac Spine (AIIS):** The ASIS, whereby the rectus femoris muscle connects, is slightly below the projection.
5. **Posterior Superior Iliac Spine (PSIS):** This is the protrusion of the iliac crest that faces posteriorly.
6. **Posterior Inferior Iliac Spine (PIIS):** It is positioned slightly below the PSIS and is abbreviated as such. The ischium is the posterior inferior hip bone part. Important landmarks include the following:

i. **Body:** It is composed of about 2/5 of the acetabulum.

ii. **Ramus:** It stretches medially from the body to attach to the pubis inferior ramus. Here are the attachments for the obturator externus, adductor magnus, and obturator internus muscles.

iii. **Ischial Tuberosity:** It is the rough, blunt protrusion of the body's inferior region that bears weight when seated. It offers a connection for the adductor Magnus and hamstring muscles (Polkowski and Clohisy, 2010).

7. **Spine:** It is situated between the lesser and larger sciatic notches in the posterior region of the body. It serves as a connection point for the sacrospinous ligament. The pubis creates the anterior

inferior hip region. It may be separated into three body sections and two rami:

i. **Body:** Externally, it constitutes approximately 1/5 of the acetabulum, while internally, it serves as an attachment point for the obturator internus muscle.
ii. **Superior Ramus:** It is located superiorly between the acetabulum and the body and serves as the muscle's attachment point.
iii. **Inferior Ramus:** It is situated posteriorly, inferiorly, and laterally to the body. Attaches the adductor brevis, adductor Magnus, and gracilis muscles.

8. **Symphysis Pubis:** It is a cartilaginous junction in the anterior midline that connects the bodies of the two pubic bones.
9. **Pubic Tubercle:** It protrudes anteriorly from the superior ramus at the pubis symphysis and serves as an attachment point for the inguinal ligament. The following structures are composed of combinations of hip bones:
   i. **Acetabulum:** Deep and cup-shaped, the acetabulum expresses with the femur. It is composed of almost equal parts of the ischium, pubis, and ileum.
   ii. **Obturator Foramen:** Obturator blood vessels and nerves enter via the foramen, which is a wide aperture enclosed by the bodies and rami of the pubis and ischium.
10. **Greater Sciatic Notch:** It is a prominent notch created into a foramen via sacrotuberous and sacrospinous ligaments right below the PIIS. Through this aperture run the piriformis muscle, sciatic nerve, and other components. The femur is the body's heaviest, strongest, and longest bone. The height of a person is about four times the size of the femur (Moore, 1985; Fuss and Bacher, 1991). It expresses with the hip bones to create the hip joint and is distinguished by the major landmarks:

i. **Head:** It is the round, cartilage-covered section that articulates with the acetabulum.
ii. **Neck:** It is the thinner area between the trochanters and the head.
iii. **Greater Trochanter:** It is a prominent projection positioned laterally among the body and neck of the femur that serves as an attachment point for the minimus and gluteus medius, as well as the majority of deep rotator muscles (Bowman et al., 2010).

11. **Lesser Trochanter:** It is a lesser projection placed posteriorly and medially immediately distal to the Greater Trochanter and serves as an attachment point for the iliopsoas muscle (Krosuri and Minor, 2003).
12. **Body:** It is the cylindrical, long region between the ends of the bones; it is also known as the shaft. It has a small anterior bow.
13. **Medial Condyle:** It is the distal medial end.
14. **Lateral Condyle:** It is the distal lateral end.
15. **Lateral Epicondyle:** This in the proximal protrusion of the lateral condyle.
16. **Medial Epicondyle:** It is the proximal extension of the medial condyle.
17. **Adductor Tubercle:** A part of the adductor Magnus muscle joins to the Adductor Tubercle, a tiny protrusion proximal to the medial epicondyle.
18. **Linea Aspera:** It is the strong longitudinal ridge or crest that extends over the majority of the posterior length.
19. **Pectineal Line:** It extends diagonally from below the lesser trochanter to the Linea Aspera. It serves to attach the adductor brevis.
20. **Patellar Surface:** It is placed anteriorly between the lateral and medial condyles. It expresses with the patella's posterior surface.
21. **Tibial Tuberosity:** It is the big proximal projection in the midline. It serves as an insertion point for the patellar tendon.

7.4. LIGAMENTS AND OTHER STRUCTURES

The hip, like other synovial joints, contains a fibrous joint capsule. It encompasses the hip joint in a cylindrical form and is robust and sturdy. It adheres proximally around the acetabular rim and distally to the femoral neck (Crowninshield et al., 1978; Byrne et al., 2010). It generates a cylindrical sleeve that completely encases the joint and the majority of the femoral neck. The capsule is strengthened by three ligaments: the ischiofemoral, pubofemoral, and iliofemoral ligaments. The iliofemoral ligament is the most essential of such ligaments. It strengthens the anterior capsule by joining the anterior inferior iliac spine proximally and crosses the joint anteriorly (Ferguson et al., 2000; Hewitt et al., 2001). It divides distally into two pieces that adhere to the intertrochanteric line of the femur. Due to its resemblance to an inverted Y, it is commonly known as the Y ligament. It is sometimes referred to as the Bigelow ligament. Its primary purpose is to prevent hyperextension (Figure 7.5) (Hewitt et al., 2002; Martin et al., 2008).

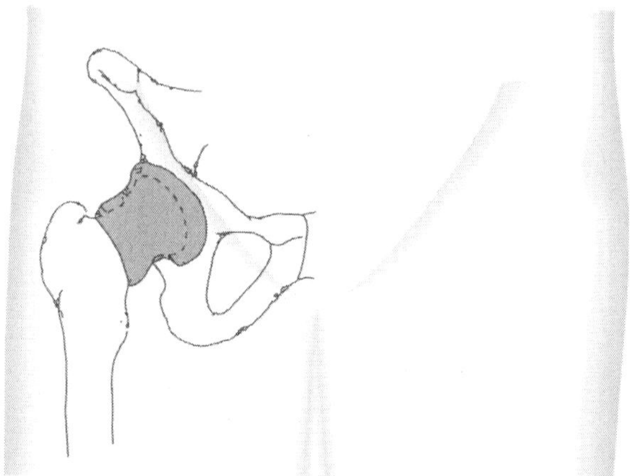

Figure 7.5. The hip joint capsule (anterior view).

Source: https://radiopaedia.org/articles/hip-joint-1.

The pubofemoral ligament runs inferiorly and medially over the hip joint. It joins to the neck of the femur by running back and down from the medial section of the acetabular rim and superior ramus of the pubis. It limits hyperextension, the same as the iliofemoral ligament. It also puts a stop to abductions (Telleria et al., 2011; Schleifenbaum et al., 2016). The posterior capsule is covered by the ischiofemoral ligament. It connects to the acetabulum's ischial section, traverses the joint in a superior and lateral orientation, and connects to the femoral neck. Hyperextension and medial rotation are limited by their fibers (Konrath et al., 1998; Van Arkel et al., 2015). All three of such ligaments connect to the acetabulum's rim and spiral over the hip joint to link to the femoral neck. This spiral connection has the combined effect of restricting mobility in a single direction (hyperextension) whereas permitting complete movement (flexion) in the other. Inflexion, such ligaments are loose, but when the hip travels into hyperextension, they become tense. By thrusting your hips forward until they are in front of your knees and shoulders, you may stand straight without engaging any muscles by relying upon the iliofemoral ligament (Inman, 1947; Clark and Haynor, 1987).

That is the foundation for a paralyzed person's standing posture after a spinal cord injury. The ligamentum teres is a minor intracapsular ligament with dubious significance. It connects to the acetabulum proximally and the fovea of the femoral head distally. Whenever the hip is semiflexed, this becomes tense throughout adduction or lateral rotation, according to certain sources. Considering the size, however, it is unlikely that it contributes considerably to the joint's strength. Another trait is that it has a blood artery that provisions the femur's head. This vascular may not provide enough blood to the cranium to keep it alive (Figure 7.6) (Pressel and Lengsfeld, 1998; Retchford et al., 2013).

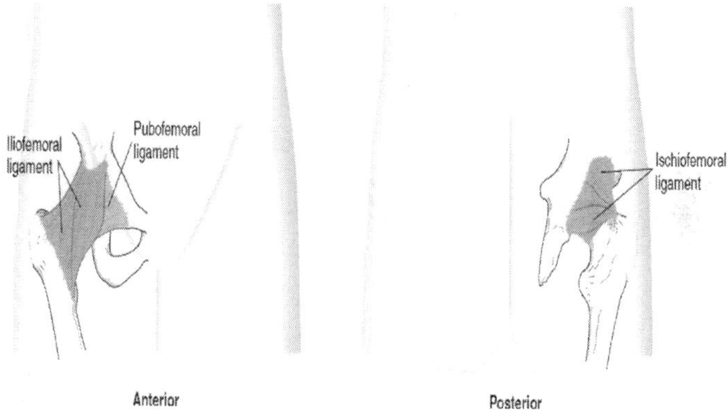

Figure 7.6. The ischiofemoral, pubofemoral, and iliofemoral ligaments provide support for the hip joint capsule.

Source: https://www.researchgate.net/figure/Extra-capsular-ligaments-reinforcing-the-hip-capsule-A-Anterior-view-B-Posterior-view_fig9_330337413.

The fibrocartilaginous acetabular labrum has been situated around the rim of the acetabulum, which contributes to an increase in the acetabulum's depth. The femoral head is encircled by the unattached end of the labrum, which helps to keep the head in its proper place in the acetabulum. While the inguinal ligament is completely useless at the hip joint, it nevertheless has to be recognized since it is present. As a separating line between the thigh and the anterior abdominal wall, it runs from the ASIS down to the pubic tubercle and extends from there. The external iliac artery and vein become the femoral artery and vein as they pass beneath the inguinal ligament (Stähelin et al., 2002; Kang et al., 2013).

The iliotibial tract or band is the tendinous component of the tensor fascia latae muscle that is extremely lengthy. It connects to the anterior section of the iliac crest and descends superficially down the lateral aspect of the thigh to the tibia. It seems unlikely that the gluteus contributes considerably to the joint's strength. It also has a blood artery that provides blood to the head of the femur. Consequently, this channel may not deliver sufficient blood to the head to stay alive. The fibrocartilaginous acetabular labrum, positioned around the rim, increases the depth of the acetabulum. The open end of the labrum encircles the femoral head and aids in maintaining the head's position in the acetabulum (Sadeghi et al., 2000; Nagano et al., 2014).

Extremely long and tendinous, the iliotibial band or tract is a component of the tensor fascia latae muscle. It is also known as the iliotibial band. It is attached to the front part of the iliac crest and travels superficially along the lateral face of the thigh until it reaches the tibia. Fibers from both the tensor fascia latae and gluteus maximus muscles join to it. Except for flexion, the final sensation of all hip joint movements is firm (soft tissue stretch) due to tension in the muscles, ligaments, and capsule. Due to contact between the anterior thigh and belly during hip flexion, the final sensation is soft (estimation of soft tissues; Figure 7.7) (Rösler and Perka, 2000; Parvaresh et al., 2019).

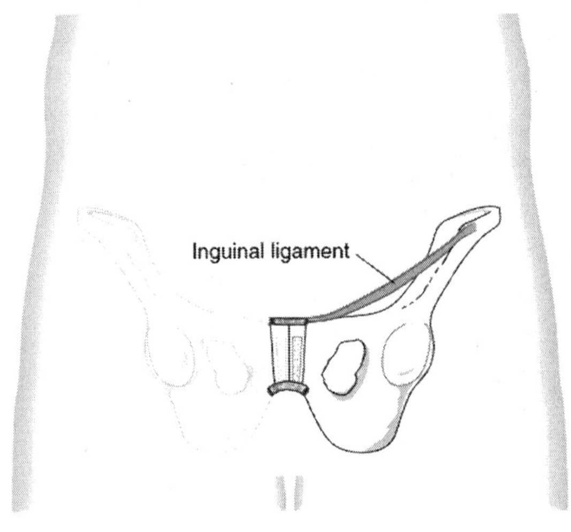

Figure 7.7. The inguinal ligament (anterior view).

Source: https://teachmeanatomy.info/abdomen/areas/inguinal-canal/.

7.5. MUSCLES OF THE HIP

Numerous parallels exist between the hip and shoulder joints. Similar to the shoulder, the hip is comprised of a set of one-joint muscles that offer the majority of control and a group of two-joint, longer muscles that supply the range of movement. Such muscles may also be categorized based on their location and, to a lesser extent, their function. For instance, anterior muscles are often flexors, lateral muscles are typically abductors, posterior muscles are typically extensors, and medial muscles are typically adductors. The hip muscles are classified by function and location The iliopsoas muscle is made up of two different muscles, each of which has its separate proximal attachment; however, the two muscles share a distal connection. The segment of the iliacus muscle originates from the iliac fossa. The component of the psoas major muscle originates from the bodies, transverse procedures, and intervertebral disks of the T12 to L5 vertebrae (McKibbin, 1970; Lazennec et al., 2004).

Collectively, such muscles connect to the lesser trochanter of the femur. The iliopsoas is the primary muscle involved in hip flexion. The part of the psoas muscle that attaches to the vertebrae helps to trunk flexion when the femur is stable. The only member of the quadriceps group to cross the hip is the rectus femoris muscle. It is attached proximally to the AIIS. This is united by the three vasti muscles and the quadriceps tendon since it descends the thigh (known as the patellar tendon). Such tendon wraps around the patella, travels through the joint of the knee, and links to the tibial tuberosity in the lower leg. The rectus femoris muscle is the primary mover in both the knee extension and the hip flexion. The sartorius muscle is the most extensive in the human body. The ASIS is where this muscle first emerges from the body. It crosses the posterior aspect of the medial knee joint after moving in a diagonal direction from the lateral to the medial and the proximal to the distal regions of the thigh. Due to its line of pull, it may abduct, bend, and laterally rotate the hip, as well as flex the knee. Consequently, this is not regarded as the driving force for any of such movements. This is most effective when all four actions are performed simultaneously. Crossing one's legs by placing one foot on the opposing knee is an illustration of this move (Moein et al., 2008; Liu et al., 2020).

The pectineus muscle is situated medial to the iliopsoas muscle and laterally to the adductor longus muscle. It begins from the superior ramus of the pubis and ends at the pectineal line of the femur. Since it extends medially and anteriorly over the hip, it facilitates hip adduction and flexion.

There have been three more hip adductors, each with a single joint. The most superficial of the three adductor muscles, the adductor longus starts from the anterior pubis surface at the tubercle and enters the middle part of the linea aspera of the femur. Due to its superficial nature, its tendon is readily palpable in the anterior-medial groin. It is essential to be able to palpate this tendon while evaluating the right fit of the quadrilateral socket of an above-knee prosthesis. It is essential for hip adduction.

The adductor brevis muscle is smaller than the other adductor muscles, as its name suggests. It sits deep with the adductor longus muscle, but superficially with the adductor Magnus's muscle. It originates from the inferior ramus of the pubis and enters above the adductor longus muscle upon this proximal linea aspera and pectineal line. It is essential for hip adduction. The adductor Magnus is the biggest and most profound of the adductor muscles. It develops from the ischial tuberosity and ramus of the ischium, as well as the pubic inferior ramus. It constitutes the bulkiest portion of the medial thigh. It is inserted all along the adductor tubercle and linea aspera. The distal connection between the adductor tubercle and linea aspera is interrupted or absent. The vein and artery of the femoral artery travel via this hole. After passing through to the posterior surface, such structures are referred to be the popliteal artery and vein, respectively (Horton and Hall, 1989; Sotereanos et al., 2006).

The adductor Magnus's muscle is a powerful hip adductor due to its size. The gracilis muscle is the only hip adductor that has two joints. The symphysis and inferior ramus of the pubis give way to a superficial and medial descent of the thigh. It curls around the medial condyle and attaches distally to the anteromedial surface of the proximal tibia after crossing the knee joint posteriorly. It aids in the flexion of the knees. The gluteus maximus muscle is a big, quadrilateral, quadrilateral, thick, one-joint muscle situated superficially on the posterior buttock. It originates in the general region of the posterior sacrum, ilium, and coccyx, and travels laterally and distally to the posterior femur, inferior to the greater trochanter, in a diagonal orientation. The iliotibial band is also where certain fibers connect. It is especially powerful in hip hyperextension, extension, and lateral rotation since it crosses the hip posteriorly in this diagonal manner.

Six tiny, deep, predominantly posterior muscles cross the hip joint horizontally and all laterally twist the hip. Since their unique connections are not functionally relevant as they all cooperate to create the same movement, they may be classed together like the deep rotator muscles. The piriformis, on either hand, is the most well-known of this group, due to its strong ties to the sciatic nerve. The hamstring muscles are a group of three muscles that cover the back of the leg. The semitendinosus, semimembranosus, and biceps femoris muscles make up this group. They all come from the same place on the ischial tuberosity.

Closer to the semitendinosus muscle on the medial side of the thigh, the semimembranosus muscle enters on the posterior surface of the medial condyle of the tibia. The distal tendon of the semitendinosus muscle is substantially longer and smaller than that of the sartorius and gracilis muscles, and it crosses the knee joint posteriorly before proceeding anteriorly to join the sartorius and gracilis muscles on the anteromedial portion of the tibia. The biceps femoris muscle has two heads and descends laterally on the posterior side of the thigh. The ischial tuberosity gives rise to the long head, as do the other two muscles, whereas the linea aspera's lateral lip gives rise to the little head. Both heads connect laterally to the head of the fibula and, by a minor slip, to the lateral condyle of the tibia, spanning the posterior part of the knee. They bend the knee because of their posterior knee covering. The hip is elongated because the big head spans the posterior hip.

The other two gluteal muscles have been positioned more laterally. The gluteus medius muscle is triangular, similar to the shoulder deltoid muscle. It joins proximally to the lateral surface of the greater trochanter and distally to the outer surface of the ilium. Since the gluteus medius muscle crosses the hip laterally, this may abduct the hip. Its anterior fibers can aid in the medial rotation of the hip via the gluteus minimus (Yi et al., 2019).

An additional critical function is performed by these two gluteal muscles, which connect to the ilium and the femur and cross the hip laterally. Whenever you stand on a single leg, the distal section of your skeleton (which is your femur) becomes significantly more stable than the proximal portion of your skeleton (which is your pelvis), and the origin moves closer to the insertion. Reversal of muscle function is another word for this shift. The opposing side of your pelvis will descend if such muscles did not activate when you stepped on one leg. When you stand on a single leg, the gluteus medius and minimus muscles engage to maintain the pelvis level and avoid the opposing side of the pelvis from sinking too far. This happens whenever you pick up a single leg, such as when walking. A "Trendelenburg gait" is caused by the weakness or loss of such muscles. If, for example, the abductors on the right side of your hip are weak, the left side of your pelvis would drop dramatically if you sit on your right leg and lift your left leg off the ground while balancing on your right leg. If you stand on your right leg and lift your left leg off the ground while doing so, you would have this problem (Lewis et al., 2007).

The tensor fascia latae muscle is a tiny muscle that is linked to a long tendon. It begins at the ASIS, crosses the hip anteriorly and laterally, and then links to the iliotibial band, along with the fascial band that extends down the lateral thigh and links to the lateral condyle of the tibia. It is an abductor of the hip, but because of its small anterior placement, its strength is likely highest when abduction and flexion are combined. To put it differently, abducting in a slight anterior direction has been the most effective (Figure 7.8) (Diamond et al., 2016).

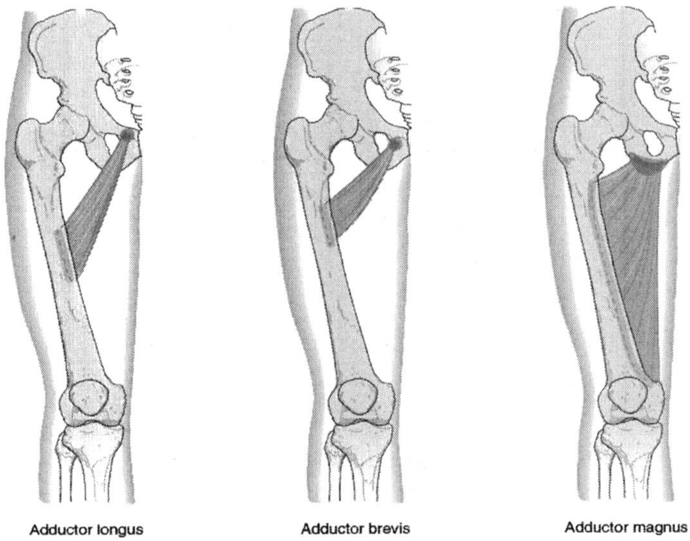

Figure 7.8. The three adductor muscles (anterior view).

Source: https://www.osmosis.org/learn/Anatomy_of_the_anterior_and_medial_thigh.

7.6. ANATOMICAL RELATIONSHIPS

The hip muscles have been divided into four groups depending upon their location. The anatomical connections of the hip muscles may be readily explained utilizing this classification by including one more factor: deep muscles vs superficial muscles.

The sartorius and the tensor fascia latae, which originate on the ASIS, have been the two superficial muscles to appear. From their shared connection, they form an inverted V. The tensor fascia latae goes somewhat lateral and down toward the knee, whereas the sartorius runs down in a medial orientation. The rectus femoris, which goes straight down toward the knee, rests among two muscles. The pectineus, iliopsoas, gracilis, and adductor longus are the muscles that move medially from the sartorius. The adductor brevis is located deep in the adductor longus in the hip, while the massive, broad adductor Magnus is located deep in the adductor brevis. The adductor Magnus is located farther distally on the thigh, deep to the adductor longus. The sartorius, the top component of the gracilis, the adductor longus, and the top half of the adductor Magnus may all be seen from front to back when looking at the hip area from the medial side, preceded by the medial hamstrings. Many of the adductor longus, as well as most of the adductor brevis and adductor Magnus, lay deep in this medial view. The gluteus maximus covers the proximal posterior hip area on the backside. The hamstring muscles are located distal to the gluteus maximus and cover the majority of the posterior thigh (Cheng et al., 2012).

The gluteus medius is deeper than the gluteus maximus and considerably more lateral than the gluteus minimus. The deep rotators are the deepest muscles; the illustration depicts five of the six deep rotators. At its proximal connection to the ischial tuberosity, the hamstring muscles lie beneath the gluteus maximus. The figure 7.8 depicts the iliotibial band laterally, the gluteus maximus posteriorly, and the tensor fascia latae anteriorly as viewed from the side of the proximal hip. The gluteus minimus and gluteus medius are located deep in such structures, respectively.

7.7. COMMON HIP PATHOLOGIES

The hip joint is subject to a wide variety of orthopedic problems that can manifest themselves at any point in one's life and have an impact on the alignment of the lower extremities. Congenital hip dislocation, also known as dysplasia, is a condition that develops when the femoral head slides upward as a result of an exceptionally shallow acetabulum. Even though it has been stretched, the joint capsule has not been damaged. Legg-Calvé-Perthes disease, also known as coxa plana, is characterized by the necrosis of the femoral head. Children between the ages of 5 and 10 years are most likely to be affected by this condition. Depending on the severity of the condition, the process of the skull dying, revascularizing, and finally remodeling may take anywhere from two to four years. During the years of rapid growth, children are at risk for developing a condition known as slipped capital femoral epiphysis. When this happens, the proximal epiphysis moves out of its usual place on the head of the femur (Yoshito et al., 2003).

The inclination angle is the angle formed in the frontal plane between the neck and the shaft of the femur, which is usually 125°. From birth to adulthood, the above angle changes. The angle can be as high as 170° at birth, however, by adulthood, it has significantly decreased. Factors like trauma, congenital deformity, or disease, on either hand, can affect the angle. A neck-shaft angle of more than 125° distinguishes Coxa valga. Since this angle is "straighter," it seems to lengthen the limb, causing the hip to be adducted while weight-bearing. Coxa vara is a disfigurement wherein the angle between the neck and the shaft is lower than 125°. Since this is "extra bent," it tries to shorten the implicated limb, causing the pelvis to drop on that side when weight-bearing.

The angle of torsion is the angle formed in the transverse plane between the neck and the shaft of the femur, with the neck and head generally rotated outward from the shaft by 15° to 25°. The femoral neck and head are superimposed on the shaft when looking down on the femur. A line through the femoral condyles, that are associated with the shaft distally, is the best way to express the shaft. The condyles rotate with the shaft. Anteversion is a rise in this angle that drives the hip joint into a more medially rotated posture. A person will walk more "toed in" as a result of this. Retroversion is defined as a reduction in torsion angle. As a result, the hip joint is forced to twist laterally, leading the person to walk more "toed out."

Osteoarthritis is a degenerative disease of the joint's articular cartilage. It can be caused by wear and tear or trauma, and it usually appears later in life. A complete joint replacement is usually used to treat it. There are two forms of the fractures of the hip: femoral and intertrochanteric neck. These are particularly frequent in the elderly and are mainly caused by falls. The fractures of the hip in younger people can be caused by high-impact trauma, like car accidents (Steultjens et al., 2000).

An overuse injury that causes lateral knee discomfort is iliotibial band syndrome. It is quite frequent in runners and bikers. The band that slips across the lateral femoral epicondyle during knee movement is thought to be the reason for this condition. Muscle tension, worn-out shoes, and jogging on uneven surfaces are all factors that contribute to it. Numerous bursae provide a friction-reducing cushion between the bone and the muscles because several muscles enter at the greater trochanter. Acute trauma or overuse might result in trochanteric bursitis. This may be seen in bicyclists or runners, or it may be induced by various circumstances that subject the greater trochanter to repetitive stress, such as a leg length discrepancy. The most frequent muscle ailment in the body is a hamstring strain, commonly known as a "pulled hamstring." Regrettably, it is a common occurrence. It might be the result of a muscular overload or an attempt to move the muscle too quickly. As a result, it is a common ailment in sports and sprinters like football, soccer, track and field, and rugby that involve bursts of speed or quick acceleration. Strains in the hamstrings may develop at each of the attachment points or anywhere along the muscle's length.

The term "hip pointer" is inaccurate since the condition affects the pelvis, not the hips. Significant bruising is the result of direct trauma to the iliac crest of the pelvis, which produced it. Although football is the sport in which it is observed most frequently, it may be seen in virtually all contact sports. It is possible that spearing the hip or the pelvis containing a helmet during tackling is the prevalent cause of this injury (Blackburn and Padua, 2008; Howard et al., 2011).

7.8. AN OVERVIEW OF MUSCLE INNERVATION

The femoral nerve controls the innervation of muscles on the thigh and hip's anterior aspect (hip flexors). The superior gluteal nerve supplies the hip abductors on the lateral side. The obturator nerve innervates the hip adductors on the medial side of the hip. The hamstring muscles, which are hip extensors and are located posteriorly, are innervated by the sciatic nerve. All generalizations have exceptions. The inferior gluteal nerve supplies innervation to the gluteus maximus, a posterior muscle.

REFERENCES

1. Blackburn, J. T., & Padua, D. A., (2008). Influence of trunk flexion on hip and knee joint kinematics during a controlled drop landing. *Clinical Biomechanics, 23*(3), 313–319.
2. Bowman, Jr. K. F., Fox, J., & Sekiya, J. K., (2010). A clinically relevant review of hip biomechanics. *Arthroscopy: The Journal of Arthroscopic & Related Surgery, 26*(8), 1118–1129.
3. Byrne, D. P., Mulhall, K. J., & Baker, J. F., (2010). Anatomy & biomechanics of the hip. *The Open Sports Medicine Journal, 4*(1).
4. Cheng, G., Gu, W., & Ge, S. R., (2012). Kinematic analysis of a 3SPS+ 1PS parallel hip joint simulator based on Rodrigues parameters. *Journal of Mechanical Science and Technology, 26*(10), 3299–3310.
5. Clark, J. M., & Haynor, D. R., (1987). Anatomy of the abductor muscles of the hip as studied by computed tomography. *The Journal of Bone and Joint Surgery; American, 69*(7), 1021–1031.
6. Crowninshield, R. D., Johnston, R. C., Andrews, J. G., & Brand, R. A., (1978). A biomechanical investigation of the human hip. *Journal of Biomechanics, 11*(1, 2), 75–85.
7. Diamond, L. E., Wrigley, T. V., Bennell, K. L., Hinman, R. S., O'Donnell, J., & Hodges, P. W., (2016). Hip joint biomechanics during gait in people with and without symptomatic femoroacetabular impingement. *Gait & Posture, 43*, 198–203.
8. Dostal, W. F., Soderberg, G. L., & Andrews, J. G., (1986). Actions of hip muscles. *Physical Therapy, 66*(3), 351–359.
9. Eyüboğlu, F., Sayaca, Ç., Çalik, M., Korkem, D., Tascilar, L. N., & Kaya, D., (2020). Kinesiology of the hip. In: *Comparative Kinesiology of the Human Body* (pp. 375–392). Academic Press.
10. Ferguson, S. J., Bryant, J. T., Ganz, R., & Ito, K., (2000). The influence of the acetabular labrum on hip joint cartilage consolidation: A poroelastic finite element model. *Journal of Biomechanics, 33*(8), 953–960.
11. Floyd, R. T., & Thompson, C. W., (2009). *Manual of Structural Kinesiology* (Vol. 16). New York, NY: McGraw-Hill.
12. Fuss, F. K., & Bacher, A., (1991). New aspects of the morphology and function of the human hip joint ligaments. *American Journal of Anatomy, 192*(1), 1–13.

13. Harput, G., (2020). Kinesiology of the knee joint. In: *Comparative Kinesiology of the Human Body* (pp. 393–410). Academic Press.
14. Harty, M., (1984). The anatomy of the hip joint. In: *Surgery of the Hip Joint* (pp. 45–74). Springer, New York, NY.
15. Hewitt, J. D., Glisson, R. R., Guilak, F., & Vail, T. P., (2002). The mechanical properties of the human hip capsule ligaments. *The Journal of Arthroplasty, 17*(1), 82–89.
16. Hewitt, J., Guilak, F., Glisson, R., & Vail, T. P., (2001). Regional material properties of the human hip joint capsule ligaments. *Journal of Orthopedic Research, 19*(3), 359–364.
17. Horton, M. G., & Hall, T. L., (1989). Quadriceps femoris muscle angle: Normal values and relationships with gender and selected skeletal measures. *Physical Therapy, 69*(11), 897–901.
18. Howard, J. S., Fazio, M. A., Mattacola, C. G., Uhl, T. L., & Jacobs, C. A., (2011). Structure, sex, and strength and knee and hip kinematics during landing. *Journal of Athletic Training, 46*(4), 376–385.
19. Inman, V. T., (1947). Functional aspects of the abductor muscles of the hip. *JBJS, 29*(3), 607–619.
20. Kaneda, K., McKean, M., Ohgi, Y. U. J. I., & Burkett, B. J., (2012). Walking and running Kinesiology in water: A review of the literature. *Journal of Fitness Research, 1*(1), 1–11.
21. Kang, S. Y., Jeon, H. S., Kwon, O., Cynn, H. S., & Choi, B., (2013). Activation of the gluteus maximus and hamstring muscles during prone hip extension with knee flexion in three hip abduction positions. *Manual Therapy, 18*(4), 303–307.
22. Konrath, G. A., Hamel, A. J., Olson, S. A., Bay, B., & Sharkey, N. A., (1998). The role of the acetabular labrum and the transverse acetabular ligament in load transmission in the hip. *JBJS, 80*(12), 1781.
23. Krosuri, S. P., & Minor, M. A., (2003). A multifunctional hybrid hip joint for improved adaptability in miniature climbing robots. In: *2003 IEEE International Conference on Robotics and Automation (Cat. No. 03CH37422)* (Vol. 1, pp. 312–317). IEEE.
24. Lazennec, J. Y., Charlot, N., Gorin, M., Roger, B., Arafati, N., Bissery, A., & Saillant, G., (2004). Hip-spine relationship: A radio-anatomical study for optimization in acetabular cup positioning. *Surgical and Radiologic Anatomy, 26*(2), 136–144.

25. Lee, B., Knabe, C., Orekhov, V., & Hong, D., (2014). Design of a human-like range of motion hip joint for humanoid robots. In: *International Design Engineering Technical Conferences and Computers and Information in Engineering Conference* (Vol. 46377, p. V05BT08A018). American Society of Mechanical Engineers.
26. Lewis, C. L., Sahrmann, S. A., & Moran, D. W., (2007). Anterior hip joint force increases with hip extension, decreased gluteal force, or decreased iliopsoas force. *Journal of Biomechanics, 40*(16), 3725–3731.
27. Liu, W., Wang, Y., Jiang, T., Chi, Y., Zhang, L., & Hua, X. S., (2020). Landmarks detection with anatomical constraints for total hip arthroplasty preoperative measurements. In: *International Conference on Medical Image Computing and Computer-Assisted Intervention* (pp. 670–679). Springer, Cham.
28. Martin, H. D., Savage, A., Braly, B. A., Palmer, I. J., Beall, D. P., & Kelly, B., (2008). The function of the hip capsular ligaments: A quantitative report. *Arthroscopy: The Journal of Arthroscopic & Related Surgery, 24*(2), 188–195.
29. McKibbin, B., (1970). Anatomical factors in the stability of the hip joint in the newborn. *The Journal of Bone and Joint Surgery; British, 52*(1), 148–159.
30. Moein, C. A., Verhofstad, M. H. J., Bleys, R. L. A. W., & Van, D. W. C., (2008). Soft tissue anatomy around the hip and its implications for choice of entry point in antegrade femoral nailing. *Clinical Anatomy: The Official Journal of the American Association of Clinical Anatomists and the British Association of Clinical Anatomists, 21*(6), 568–574.
31. Morcelli, M. H., LaRoche, D. P., Crozara, L. F., Marques, N. R., Hallal, C. Z., Rossi, D. M., & Navega, M. T., (2016). Neuromuscular performance in the hip joint of elderly fallers and non-fallers. *Aging Clinical and Experimental Research, 28*(3), 443–450.
32. Murray, M. P., Brewer, B. J., Gore, D. R., & Zuege, R. C., (1975). Kinesiology after McKee-Farrar total hip replacement. A two-year follow-up of one hundred cases. *The Journal of Bone and Joint Surgery; American, 57*(3), 337–342.
33. Nagano, Y., Higashihara, A., Takahashi, K., & Fukubayashi, T., (2014). Mechanics of the muscles crossing the hip joint during sprint running. *Journal of Sports Sciences, 32*(18), 1722–1728.

34. Neumann, D. A., (2010). Kinesiology of the hip: A focus on muscular actions. *Journal of Orthopedic & Sports Physical Therapy, 40*(2), 82–94.

35. Nonaka, H., Mita, K., Watakabe, M., Akataki, K., Suzuki, N., Okuwa, T., & Yabe, K., (2002). Age-related changes in the interactive mobility of the hip and knee joints: A geometrical analysis. *Gait & Posture, 15*(3), 236–243.

36. Parvaresh, K. C., Chang, C., Patel, A., Lieber, R. L., Ball, S. T., & Ward, S. R., (2019). Architecture of the short external rotator muscles of the hip. *BMC Musculoskeletal Disorders, 20*(1), 1–6.

37. Polkowski, G. G., & Clohisy, J. C., (2010). Hip biomechanics. *Sports Medicine and Arthroscopy Review, 18*(2), 56–62.

38. Pressel, T., & Lengsfeld, M., (1998). Functions of hip joint muscles. *Medical Engineering & Physics, 20*(1), 50–56.

39. Rallis, I., Langis, A., Georgoulas, I., Voulodimos, A., Doulamis, N., & Doulamis, A., (2018). An embodied learning game using Kinect and labanotation for analysis and visualization of dance Kinesiology. In: *2018 10th International Conference on Virtual Worlds and Games for Serious Applications (VS-Games)* (pp. 1–8). IEEE.

40. Retchford, T. H., Crossley, K. M., Grimaldi, A., Kemp, J. L., & Cowan, S. M., (2013). Can local muscles augment stability in the hip? A narrative literature review. *J. Musculoskelet Neuronal Interact., 13*(1), 1–12.

41. Rösler, J., & Perka, C., (2000). The effect of anatomical positional relationships on kinetic parameters after total hip replacement. *International Orthopedics, 24*(1), 23–27.

42. Rosner, A. L., & Cuthbert, S. C., (2012). Applied Kinesiology: Distinctions in its definition and interpretation. *Journal of Bodywork and Movement Therapies, 16*(4), 464–487.

43. Rutherford, D. J., Moreside, J., & Wong, I., (2015). Hip joint motion and gluteal muscle activation differences between healthy controls and those with varying degrees of hip osteoarthritis during walking. *Journal of Electromyography and Kinesiology, 25*(6), 944–950.

44. Sadeghi, H., Prince, F., Sadeghi, S., & Labelle, H., (2000). Principal component analysis of the power developed in the flexion/extension muscles of the hip in able-bodied gait. *Medical Engineering & Physics, 22*(10), 703–710.

45. Schleifenbaum, S., Prietzel, T., Hädrich, C., Möbius, R., Sichting, F., & Hammer, N., (2016). Tensile properties of the hip joint ligaments are largely variable and age-dependent–An in-vitro analysis in an age range of 14–93 years. *Journal of Biomec Hanics, 49*(14), 3437–3443.
46. Sotereanos, N. G., Miller, M. C., Smith, B., Hube, R., Sewecke, J. J., & Wohlrab, D., (2006). Using intraoperative pelvic landmarks for acetabular component placement in total hip arthroplasty. *The Journal of Arthroplasty, 21*(6), 832–840.
47. Stähelin, T., Vienne, P., & Hersche, O., (2002). Failure of reinserted short external rotator muscles after total hip arthroplasty. *The Journal of Arthroplasty, 17*(5), 604–607.
48. Steultjens, M. P. M., Dekker, J. V., Van, B. M. E., Oostendorp, R. A. B., & Bijlsma, J. W. J., (2000). Range of joint motion and disability in patients with osteoarthritis of the knee or hip. *Rheumatology, 39*(9), 955–961.
49. Telleria, J. J., Lindsey, D. P., Giori, N. J., & Safran, M. R., (2011). An anatomic arthroscopic description of the hip capsular ligaments for the hip arthroscopist. *Arthroscopy: The Journal of Arthroscopic & Related Surgery, 27*(5), 628–636.
50. Van, A. R. J., Amis, A. A., Cobb, J. P., & Jeffers, J. R. T., (2015). The capsular ligaments provide more hip rotational restraint than the acetabular labrum and the ligamentum teres: An experimental study. *The Bone & Joint Journal, 97*(4), 484–491.
51. Yi, L. H., Li, R., Zhu, Z. Y., Bai, C. W., Tang, J. L., Zhao, F. C., & Guo, K. J., (2019). Anatomical study based on 3D-CT image reconstruction of the hip rotation center and femoral offset in a Chinese population: Preoperative implications in total hip arthroplasty. *Surgical and Radiologic Anatomy, 41*(1), 117–124.
52. Yoshito, O. K. H., Naoki, S. A. H., & Nobuhiko, S. K., (2003). 4-dimensional computer-based motion simulation after total hip arthroplasty. *Medicine Meets Virtual Reality 11: NextMed: Health Horizon, 94*, 251.

CHAPTER 8

KINESIOLOGY OF ANKLE JOINT AND FOOT

CONTENTS

8.1. Introduction 194
8.2. Functional Characteristics of the Foot 197
8.3. Joints and their Motions 198
8.4. Ligaments and Other Structures 204
8.5. Muscles of the Foot and Ankle 206
8.6. Anatomical Associations 210
8.7. Summary of Muscle Innervation 212
8.8. Usual Ankle Pathologies 212
References 215

8.1. INTRODUCTION

The tibia and fibula are the bones that make up the leg which is the part of the lower end that runs from the knee to the ankle. A robust interosseous membrane connects these two bones and gives muscles more surface area to bind to (Figure 8.1) (Kwon and Park, 2012).

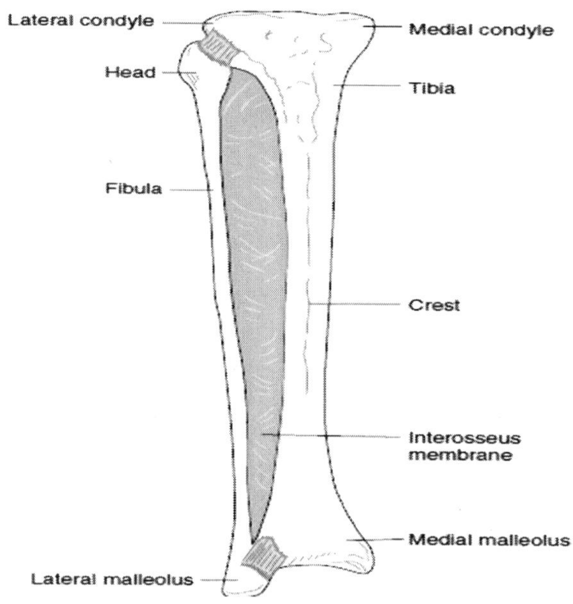

Figure 8.1. Interosseous membrane and leg bones.

Source: https://clinicalgate.com/leg/.

The larger bone out of the two, the tibia, is the leg's only bone which bears the weight. The apex (crest) of the tibia is positioned anteriorly and is triangular in shape. The fibula, which is long and narrow, is aligned with the tibia's posterior surface. This creates a passage lateral to the tibia, with the interosseous membrane as the floor, allowing multiple muscles to attach without changing the leg's morphology. The following are tibia landmarks related to the ankle (Figure 8.2) (Kim et al., 2015):

1. **Medial Condyle:** The middle adjacent end.
2. **Lateral Condyle:** The proximal end of the side.
3. **Crest:** The most visible and anterior of the three borders.
4. **Medial Malleolus:** The expanded medial distal surface (Lee and Lee, 2015).

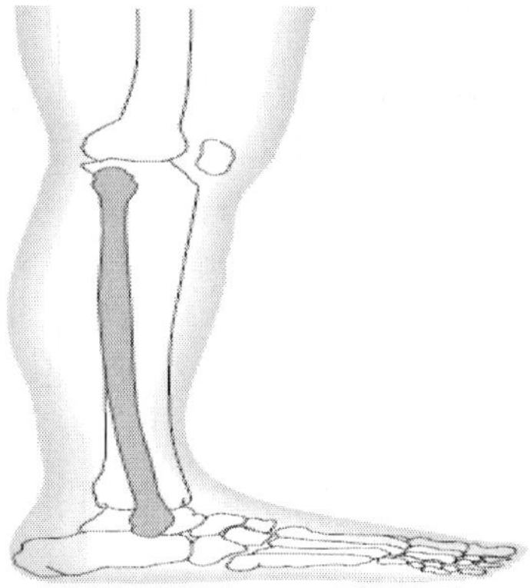

Figure 8.2. Lateral view of right leg. Notice the fibula's posterior position.

Source: https://quizlet.com/238937511/tibia-and-fibula-lateral-view-diagram/.

Following are the fibula's landmarks (Floyd and Thompson, 2009):

i. **Head:** Expanded adjacent end.
ii. **Lateral Malleolus:** Distal end expanded.

The metatarsals, phalanges, and tarsals are the bones of the foot. The following are the seven tarsal bones and their landmarks (Jackson et al., 2016):

1. **Calcaneus:** The biggest and posterior-most tarsal bone.
2. **Calcaneal Tuberosity:** Projection on the calcaneus's posterior lower surface.
3. **Sustentaculum Tali:** It is the superior medial section of the calcaneus which projects out from the remainder of the calcaneus and supports the talus's medial side. Three tendons wrap around this protrusion, switching from the back leg to the plantar foot (Glassow, 1966).
4. **Talus:** It is the 2nd largest tarsal, sitting on the calcaneus.
5. **Navicular:** On the middle side, in front of the talus and close to the three cuneiforms.
6. **Tuberosity of Navicular:** Projection of the navicular on the medial side; easily visible on the middle border of the foot.
7. **Cuboid:** Adjacent to the 4th and 5th metatarsals and away from the calcaneus on the sideways of the foot.

8.1.1. Cuneiforms

The first through third metatarsals are three in number and are numbered from medial to lateral in accordance with the metatarsals. The first of the three is the largest. Starting medially, the metatarsals are counted one through five. The 1st and 5th metatarsals are normally the bones which bear weight, whereas the 2nd, 3rd, and 4th metatarsals bear no weight. We have a habit of standing in a triangle. The weight is carried from the calcaneus to the 1st and 5th metatarsal heads. The metatarsals have the following distinguishing characteristics and landmarks (Kim and Shin, 2017):

- **Base:** Each metatarsal's proximal end.
- **Head:** Each bone's distal end.
- **First:** On the middle side of the foot, the shortest and thickest metatarsal. The initial cuneiform articulation.
- **Second:** It articulates with the 2nd cuneiform and is the longest.
- **Third:** The 3rd cuneiform articulates.
- **Fourth:** It articulates with the cuboid together with the 5th metatarsal.
- **Fifth:** Possesses a noticeable tuberosity on the sideways of its base. The makeup of the foot's phalanges is identical to those of the hand. The great toe, the 1st digit, has adjacent and isolated phalanx but no intermediate phalanx. Each of the 2nd through 5th digits, often known as the four smaller toes, is composed of middle, proximal, and distal phalanx (Figure 8.3) (Long et al., 2017).

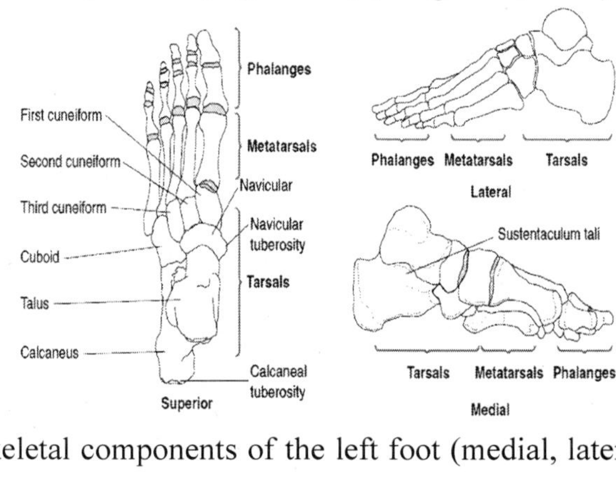

Figure 8.3. Skeletal components of the left foot (medial, lateral, and superior views).

Source: https://www.coursehero.com/study-guides/nemcc-ap/bones-of-the-lower-limb/.

8.2. FUNCTIONAL CHARACTERISTICS OF THE FOOT

The foot is split into three main sections. The calcaneus and talus make up the hind foot. The hind foot is the 1st section of the foot to make contact with the ground throughout the gait cycle, affecting the movement and function of the other two. The cuboid, the navicular, and the three cuneiform bones make up the mid foot. As it conveys movement from the hind foot to the forefoot, the mechanics of this region of the foot give mobility and stability. The five metatarsals, along with all the phalanges, make up the forefoot. This portion of the foot adjusts to the ground's level. During the stance phase, it is the foot's last part to make contact with the ground (Figure 8.4) (Sarrafian, 1987).

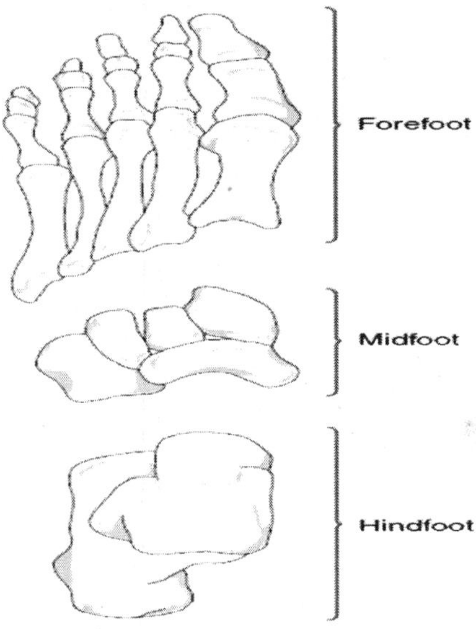

Figure 8.4. The foot's functional areas (superior view).

Source: https://www.kenhub.com/en/library/anatomy/ankle-and-foot-anatomy.

The ankle joint and foot provide three primary purposes (Gabbard and Hart, 1996). As the foot's heel touches the ground at the start of the stance phase, the foot acts as a shock absorber, responding to the unevenness of the ground and giving a firm base of support to move the body forward (Fong et al., 2008).

8.3. JOINTS AND THEIR MOTIONS

8.3.1. Motions of Ankle

Because there is not universal agreement among authors, ankle joint and foot motions must be defined. Plantar flexion is motion towards the foot's plantar surface, while dorsiflexion is movement towards the leg's anterior surface from the foot's dorsal surface. These movements take place in the sagittal plane, around the frontal axis. The terminologies extension and flexion should not be used because they have contradictory definitions. Plantar flexion, for example, is functionally equivalent to extension in that it is part of the knee, hip, and ankle extension movement. However, plantar flexion is not a real flexion anatomically since there is no measurement of two segments (Rougier and Caron, 2000).

Eversion and inversion refer to the movement in a frontal plane around the sagittal axis. Inversion is the inward turning of the forefoot caused by increasing the foot's medial edge. The opposite action, eversion, is the rising of the foot's lateral boundary, which turns the forefoot outward. Abduction and adduction are terms used to describe movement in the transverse plane. These movements occur largely in the forefoot and are associated with inversion and eversion, accordingly (Figure 8.5) (Close, 1956).

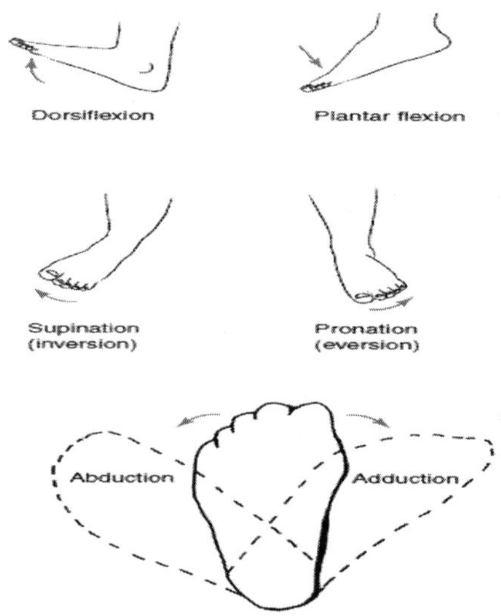

Figure 8.5. Ankle joint and motions of the foot.

Source: https://www.researchgate.net/figure/Ankle-joint-movements_fig1_282073553.

Clinicians have begun to use the terms pronation and supination to characterize ankle joint and foot movements in recent years. Pronation is a combination of abduction, dorsiflexion, and eversion, while supination is a blend of plantar flexion, adduction, and inversion. Valgus and varus should be defined to avoid further misunderstanding. These terminologies are most typically used to describe a certain position, usually one that is aberrant. The position of the distal section distant from the midline is referred to as valgus. Varus, on the other hand, is a posture in which the distal part is closer to the midline (Oatis, 1988).

The distal section of the calcaneus is thus tilted away from the midline, resulting in calcaneal valgus. These phrases will not be utilized here because the emphasis is on motion rather than position. In summary, the terms plantar flexion, supination (a blend of plantar flexion, forefoot adduction, and inversion), dorsiflexion, and pronation (a blend of dorsiflexion, forefoot abduction, and eversion) are often utilized by doctors to describe motions of the ankle and foot. Figure 8.5 illustrates these movements. However, inversion and eversion are employed instead of supination and pronation when discussing muscle motion (Kipp and Palmieri-Smith, 2013).

The tibiofibular joints are two minor joints that are not part of the main ankle joint but have a little role in the normal job of the ankle. The articulation between the fibula's head and the posterior lateral part of the proximal tibia is known as the superior tibiofibular joint. It is a plane joint that permits the fibula to glid and rotates on the tibia to a limited extent. It possesses a joint capsule because it is a synovial joint (Hargrove et al., 2011). The capsule is reinforced by ligaments, and the joint works to relieve torsional stresses at the ankle joint. The syndesmosis between the convex distal fibula and the concave distal tibia forms the inferior tibiofibular joint. There is no joint capsule because it is not a synovial joint. Fibrous tissue, on the other hand, splits the bones and the ligaments that hold the joint together. A robust union at the ankle joint is essential for the joint's strength. The ligaments that hold the inferior tibiofibular joint together permit a small amount of flexibility to accommodate talus motion (Loudon and Bell, 1996).

8.3.2. Ankle Joints

The true ankle joint is composed of the distal tibia, which lies on the talus, with the tibia's medial malleolus fitting down around the talus's medial surface, and the fibula's lateral malleolus fitting down around the talus's lateral aspect. Tenon and mortise joints are common carpentry terms for this sort of junction. A mortise is a notch carved in a wood piece to take a shaped protruding piece. As a result, the mortise would be the malleoli of the tibia and fibula, and the tenon would be the talus. This joint connects the leg with the foot and controls the bulk of foot motion compared to the leg (Astephen et al., 2008).

The ankle is a uniaxial hinge joint made up of articulation between the tibia's distal end and the medial malleolus, as well as the fibula's lateral malleolus and the talus. Plantar flexion is limited to 30° to 50°, and dorsiflexion is limited to 20°. The ankle is generally in a neutral posture in the anatomical position. Triplanar motion is defined as movement around an unevenly oriented axis that goes through all three planes since the axis of rotation is at an angle (Figure 8.6) (Roaas and Andersson, 1982).

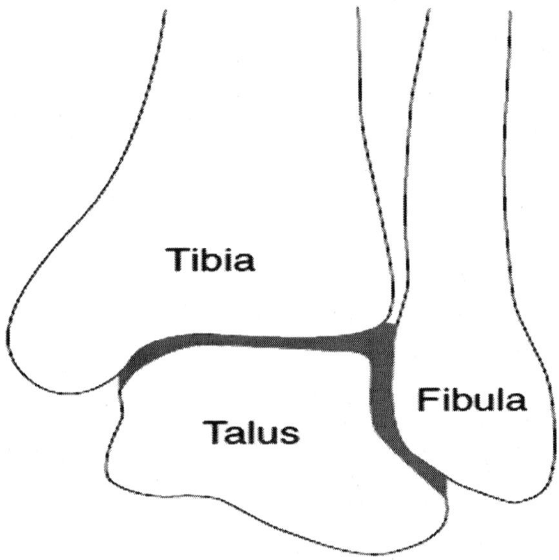

Figure 8.6. Posterior view of ankle joint.

Source: https://www.lecturio.com/concepts/ankle-joint/.

The lateral malleolus extends further distally and resides posteriorly as compared to the medial malleolus along this axis. Keep the tips of your forefingers at the distal ends of your left ankle's malleoli to imagine this position. The forefinger on the lateral malleolus side is more anterior when viewed from above. The forefinger is more distal when observed from the front. Consider the fingers to be a straight rod that passes through the joint (Zhang et al., 2002). The forefingers do not line up in a straight line from side to side. The left finger has a slight posterior and inferior position, whereas the right finger has a slight anterior and superior position. The ankle joint's axis is basically this. It leans 8° away from the transverse plane, 82° away from the sagittal plane, and 20–30° away from the frontal plane. The foot slightly moves out, not only comes up during ankle dorsiflexion. The foot moves down and in during ankle plantar flexion (Braunstein et al., 2010). The joint axis's angle leads the foot to adduct during plantar flexion and abduct during dorsiflexion in an open kinetic chain with the leg fixed and the foot free to move. Closed chains have the opposite effect (Sereno and Arcucci, 1990).

The foot is immobile while the leg moves across the ground. Throughout dorsiflexion, the leg rotates medially on the foot. When the foot remains stationary and the leg moves over it, the joint axis angle triggers the leg to spin medially over the foot. During plantar flexion of the ankle, the leg rotates horizontally on the foot. This rotation is permitted due to the small range of motion at the tibiofibular joints (Schweitzer et al., 1994). It is a supplementary motion comparable to the movement of the thumb's CMC joint. This movement cannot be performed in an open chain. During ankle dorsiflexion, the convex talus glides laterally on the concave tibia, whereas it glides anteriorly during ankle plantar flexion. Dorsiflexion and plantar flexion result in a firm sensation and are characterized as soft tissue stretch. This is because the joint capsule, tendons, and ligaments are under tension (Chubinskaya et al., 1999).

The talocalcaneal, or subtalar, joint is created by the talus' inferior surface articulating with the calcaneus' superior surface. It is a one-degree-of-freedom plane synovial joint. Inversion and eversion take place around an oblique axis. The anterior sides of the calcaneus and talus articulate with the posterior surfaces of the cuboid and navicular, respectively, to form the transverse tarsal joint (Puno et al., 1991). The cuboid and the navicular move relatively little, even though they are right adjacent to each other. In inversion and eversion, the transverse tarsal joint connects the hindfoot and forefoot. They are combinations of movements since the movements of these two joints are on an oblique axis. The transverse tarsal and subtalar joints can't be separated functionally. Inversion or eversion will be utilized to define motions at the subtalar and transtarsal joints. A mixture of supination, plantar flexion, and adduction will be used in inversion, whereas pronation, dorsiflexion, and abduction will be used in eversion (Hendren and Beeson, 2009).

As a result, when the ankle moves in dorsiflexion and plantar flexion, the talocrural joint is largely involved. The transverse tarsal and subtalar joints are the primary joints involved in eversion and inversion of the ankle. All of these joints work together to permit the foot to take practically any posture in space. This helps the foot adjust to irregular surfaces, such as those encountered when walking on rough ground. Consider how many different foot postures are required when hiking on rocks at the seashore or in the mountains (Banerjee and Agarwal, 1998).

8.3.3. Foot Joints

The metatarsal heads articulate with the proximal phalanges in the metatarsophalangeal (MTP) joints. There are five joints that allow extension, flexion, hyperextension, adduction, and abduction, similar to the metacarpophalangeal (MCP) joints of the hand. The first MTP joint is a lot more movable than the second. It has 45° of flexion and extension, as well as 90° of hyperextension. Only roughly 45° of hyperextension are allowed at the second through fifth MTP joints. During the toe-off phase of walking, hyperextension is crucial (Figure 8.7) (Eerdekens et al., 2019).

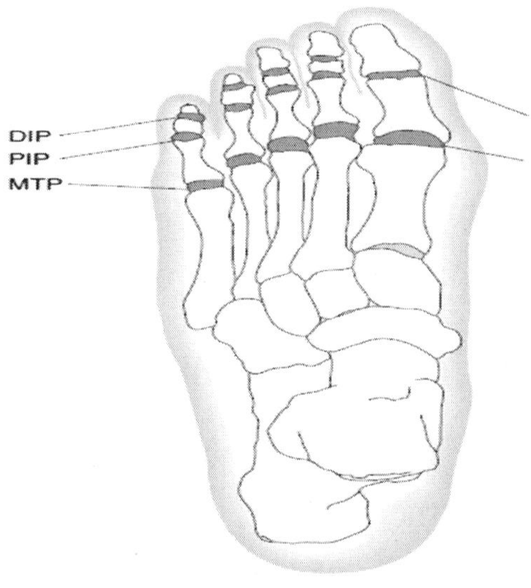

Figure 8.7. Phalanges of the foot's joints.

Source: https://www.msdmanuals.com/professional/musculoskeletal-and-connective-tissue-disorders/foot-and-ankle-disorders/metatarsophalangeal-joint-pain.

The second toe serves as a reference point for adduction and abduction. The 2nd toe, like the middle finger, abducts in both directions but only adducts as the return motion from abduction. Each of the lesser toes possesses an adjacent interphalangeal (IP) and distal interphalangeal (DIP) joint, similar to the hand. Since the foot needs less dexterity than the hand, these joints are not as important individually. The proximal and distal phalanxes of the great toe are present, but there is no intermediate phalanx. As a result, it contains just one phalangeal joint, the IP joint, similar to the thumb (Figure 8.8) (Huson, 1987).

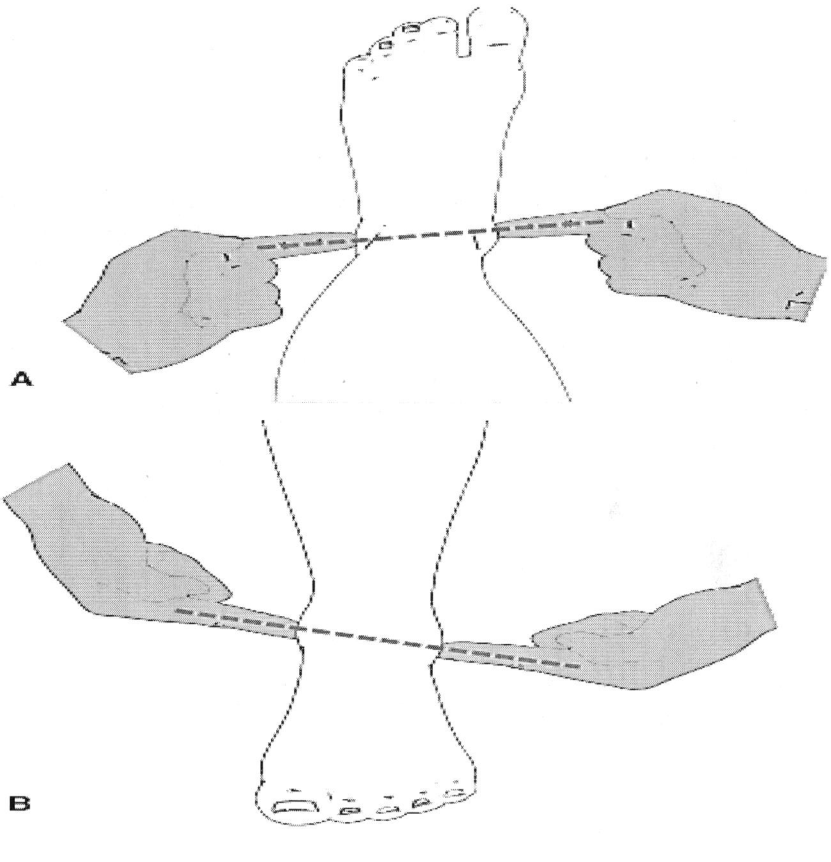

A

B

Figure 8.8. The kinematic axes of the ankle joint. (a) Superior view; (b) anterior view.

Source: https://maplespub.com/article/Joint-Pressure-Volume-and-Alignment-in-Development-of-AOA-Indications-for-Orthobiologics-and-Surgeons.

8.4. LIGAMENTS AND OTHER STRUCTURES

A joint capsule surrounds the ankle joint, which is a synovial joint. The joint capsule is very thin anteriorly and posteriorly, but it is strengthened on both sides by collateral ligaments. These collateral ligaments are basically collections of ligaments. The medial collateral ligament is a triangular deltoid ligament with an apex around the tip of the medial malleolus. Its wide base expands out in four segments to connect to the navicular, calcaneus, and talus. The anterior fibers are connected to the navicular. The middle fibers run down directly to the calcaneus sustentaculum tali. The posterior fibers extend all the way back to the talus (Woo et al., 1999).

Because they are so close to the tibionavicular section, the deep fibers are barely visible from the medial side. The deltoid ligament assists in maintaining the medial longitudinal arch by strengthening the ankle joint's medial side, holding the calcaneus and navicular against the talus, and strengthening the medial side of the ankle joint (Amis et al., 2003).

A collection of three ligaments on the ankle joint's lateral side joint are together known as the lateral ligament. The lateral ligament connects the lateral malleolus to the talus and calcaneus in three segments. The anterior talofibular ligament is relatively weak and connects the lateral malleolus to the talus. The posterior talofibular ligament connects the lateral malleolus to the talus and runs almost horizontally in the back. The calcaneofibular ligament is located in the center and connects the malleolus to the calcaneus. Several more ligaments connect the tarsals to one another, the metatarsals, etc. The bones to which they join are usually given names (Amiel et al., 1983).

Since the foot is the contact point with the ground most of the time, it should be able to take a lot of shock, react to variations in terrain, and move the body forward. The bones of the foot are organized in arches to allow these motions to take place. We stand on a triangle, which spreads weight-bearing from the calcaneus base to the first and fifth metatarsal heads. Two arches (lateral and medial longitudinal) are at 90° to the third arch between these three places (Figure 8.9) (Amis et al., 2006).

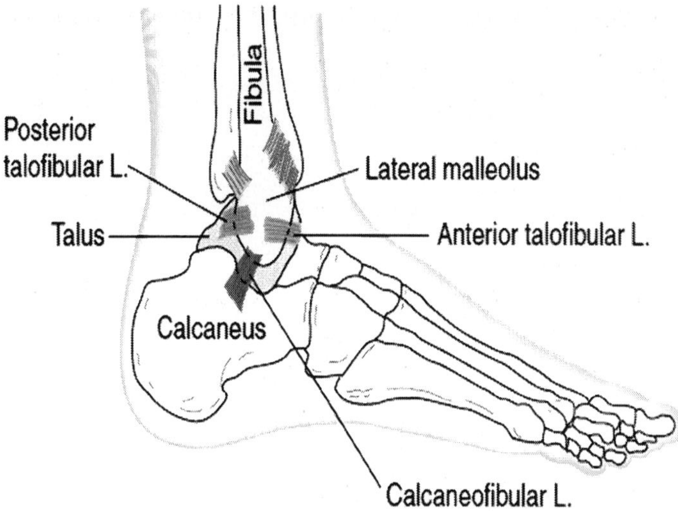

Figure 8.9. Ligaments of the right ankle's lateral aspect.

Source: https://www.nlhealthhub.com.au/what-is-a-lateral-ankle-sprain-or-rolled-ankle/.

The medial longitudinal arch moves from the calcaneus anteriorly via the navicular, the talus, and three cuneiforms anteriorly to the first three metatarsals, forming the medial boundary of the foot. The talus is located at the arch's top and is known as the keystone since it bears the body's weight. The keystone, which is usually the center or uppermost section of an arch, is an important part. When the arch bears weight, it depresses slightly and flinches when the weight is released. It never straightens or touches the ground normally (Poynton et al., 1997).

The lateral longitudinal arch travels from the calcaneus to the 4th and 5th metatarsals, passing through the cuboid. During weight-bearing, it usually rests on the ground. The transverse arch connects the three cuneiforms to the cuboid by running from side to side. The keystone of this arch is the second cuneiform. The profile of the bones and their relationship to one another, the plantar ligaments and fascia, and the muscles all work together to keep these three arches in place. Perhaps the most essential features are the ligaments and fascia. The spring ligament connects the calcaneus to the navicular and moves forward. It is wide and short, and it is crucial since it supports the longitudinal arch's medial side (DeLancey, 1989).

The long planar ligament is more superficial than the spring ligament, which is the longest of the tarsal ligaments. It joins posteriorly to the calcaneus and extends anteriorly to the cuboid and bases of the 3rd, 4th, and 5th metatarsals. It is the principal adjacent longitudinal arch support. The short plantar ligament supports the long plantar ligament and connects the calcaneus to the cuboid (Jones et al., 1999). It resides primarily near the plantar ligament. The superficially situated plantar fascia extends from the calcaneus to the adjacent phalanges and supports both longitudinal arches. It works as a tie-rod, preventing the posterior segments from detaching from the anterior segment. This plantar fascia enhances the foot and arch's stability during walking and weight-bearing (Rumian et al., 2007).

Muscles, namely the evertors and invertors of the foot, also support the arches. The flexor hallucis longus, tibialis posterior, and the flexor digitorum longus muscles all span the ankle's medial side, passing under the calcaneus sustentaculum tali. As a result, they provide support to the foot's medial side (Gupte et al., 2003). The medial longitudinal arch is supported by the flexor hallucis longus and flexor digitorum longus muscles, which bridge it. The transverse and lateral longitudinal arches are supported by the peroneus longus muscle, which extends the foot from lateral to medial. Because they are involved in every move, the intrinsic muscles offer more support as compared to the extrinsic muscles. The overall muscle support for the arches, on the other hand, is expected to bear about 15–20% of the entire stress on the arches (Akdemir, 2010).

8.5. MUSCLES OF THE FOOT AND ANKLE

8.5.1. Extrinsic Muscles

Intrinsic and extrinsic muscles exist in the foot and ankle, just as they do in the wrist and hand. Extrinsic muscles come from the leg, while intrinsic muscles come from the tarsal bones. Extrinsic leg muscles are discovered in four anatomical locations and are discovered in groups of three or mixtures of three. Those four anatomical sections correspond to the four chambers of the leg, which are separated by thick fascia. A set of muscles with a shared function exists within each compartment and are (Kurihara et al., 2014):

- Superficial posterior;
- Deep posterior;
- Anterior;
- Lateral groups/compartments.

All traverse the ankle joint and have adjacent connections on the tibia, femur, or fibula (Hunt et al., 2001). The gastrocnemius, plantaris, and soleus muscles make up the superficial posterior group. The gastrocnemius muscle connects the knee and ankle and has two joints. It is a powerful plantar flexor in the ankle. It joins to the posterior side of the femur's lateral and medial and condyles with two heads. It creates a shared Achilles tendon (commonly known as heel cord) with the soleus muscle and joins to the posterior surface of the calcaneus after descending the posterior leg superficially. Even though its primary role is at the ankle, it also spans the back of the knee and plays an important role there (Gheitasi et al., 2022). Soleus muscle is a big, one-joint muscle that lies deep within the gastrocnemius. It starts on the posterior tibia and fibula and runs down the back of the leg, merging with the gastrocnemius muscle to create the massive, robust Achilles tendon that attaches to the posterior calcaneus. The soleus muscle's only purpose is to plantar flex the ankle because it spans the midline of the ankle. The triceps surae muscle, which literally means "three-headed calf," is made up of the two heads of the soleus and gastrocnemius muscles (Soysa et al., 2012).

The plantaris muscle is a 2-joint, long, thin muscle that serves no important purpose. It begins on the sideways epicondyle of the femur's posterior surface, extents the posterior leg medially, and merges with the soleus and gastrocnemius muscles in the Achilles tendon. It must plantar flex the ankle and flex the knee in theory. Conversely, due to its small size in comparison to the main drivers of those acts, it is only helpful (Donatelli, 1985).

The tibialis posterior, flexor digitorum longus, and flexor hallucis longus muscles make form the deep posterior group. They all end in the foot and join to the posterior tibia or fibula. They can plantar flex the ankle since they all cross it posteriorly. However, due to their small size in comparison to the gastrocnemius and soleus muscles, their part in ankle plantar flexion is primarily supportive. The deepest-lying posterior muscle is the tibialis posterior. The interosseous membrane and neighboring sections of the tibia and fibula serve as its proximal attachment points (McKeon et al., 2015). It runs down the back of the leg, looping around the medial malleolus before attaching to the navicular with fibrous growths to the three cuneiforms, the cuboid, the calcaneus sustentaculum tali, and the bases of the 2nd through 4th metatarsals. The tibialis posterior muscle can invert and plantar flex the ankle since it crosses it medially and posteriorly. Due to its small size in comparison to some other plantar flexors, it is solely useful in plantar flexion (Guillén-Rogel et al., 2017).

Originating from the interosseous membrane and posterior fibula, the flexor hallucis longus muscle is mostly located on the lateral aspect of the leg. It descends posteriorly from the tibia, encircles the medial malleolus via a channel in the posterior talus, and passes beneath the calcaneus's talus sustentaculum. This muscle goes down the foot via two heads of the flexor hallucis brevis muscle to join at the base of the great toe's distal phalanx. This distal attachment is comparable to the hand's superficialis and flexor digitorum profundus muscles. The flexor hallucis longus muscle twists the great toe and aids with inversion as well as, to a slighter extent, ankle plantar flexion (Nogueira et al., 2015).

The flexor digitorum longus muscle originates from the posterior tibia and is mostly seen on the medial side of the leg. It moves down the foot, dividing into four tendons and inserts into the distal phalanx of the 2nd through 5th toes after looping around the medial malleolus (Angin et al., 2018). In the same way that the flexor digitorum profundus muscle passes through the break in the flexor digitorum brevis tendon, this muscle passes through the gap in the flexor digitorum superficialis muscle in the hand. It helps with plantar flexion and ankle inversion by flexing the four lesser toes. The deep posterior muscles' interactions are fascinating because they intersect and entangle from their middle to distal attachments. The tibialis posterior muscle is in the center of these three muscles at their origins (Kirby, 2017).

The flexor digitorum longus is located in the midst of the loop nearby the medial malleolus. The flexor hallucis longus is at the midst of their enclosures. The flexor digitorum longus is on the other side of the body from where it existed at the start. This dynamic connection feature adds strength, similar to how a braided rope is stronger as compared to the rope in which an individual strands run parallel to one another (Scarton et al., 2018).

The tibialis anterior, extensor digitorum longus, and extensor hallucis muscles make form the anterior muscular group. They all connect to the anterior lateral leg proximally and pass the ankle anteriorly. The tibialis anterior muscle arises from the sideways of the tibia and the interosseous membrane, inclines the leg, and inserts medially on the 1st cuneiform and the base of the 1st metatarsal. It accounts for the majority of the mass of the anterior adjacent leg. The tibialis anterior muscle dorsiflexes and inverts the ankle by spanning it anteriorly and medially. The extensor hallucis longus muscle, which originates on the interosseous membrane and fibula and enclosures into the base of the great toe's isolated phalanx, is a slender muscle that lies deep to and between the extensor digitorum longus and tibialis anterior muscles (Cheung and Zhang, 2006).

Its main function is to stretch the great toe, although it also helps with dorsiflexion and inversion of the ankle. The most lateral out of the anterior muscles is the extensor digitorum longus. It attaches to the interosseous membrane, the adjacent condyle of the tibia, and the majority of the anterior fibula. It attaches to the isolated phalanx of the four lesser toes as it descends the leg (Figure 8.10). The extensor digitorum longus muscle is responsible for extending the second through fifth toes, as well as dorsiflexing the ankle. Since it passes the joint via the inside of that axis, it has no inversion/eversion function (Klaue, 1997).

The peroneus brevis, peroneus tertius, and peroneus longus muscles make up the lateral group of muscles. All of them start around the fibula and move laterally to the foot. Two of them pass the joint of ankle posteriorly, whereas the other crosses it anteriorly. The peroneus longus is the peroneal muscle that is the most superficial. It descends the adjacent leg and circles behind the lateral malleolus with the peroneus brevis muscle, originating from the proximal end of the interosseous membrane and fibula (Park and Park, 2018). The peroneus longus muscle travels deep at this point, traversing the foot laterally from the sideways to the medial side and penetrating into the first cuneiform and first metatarsal plantar surfaces. This distal connection is quite near to the tibialis anterior muscle's connection. Since the peroneus longus muscle lowers the leg sideways before passing the foot medially to reach the tibialis anterior muscle, the tibialis anterior and peroneus longus muscles are frequently mentioned as the stirrup of the foot. The tibialis anterior muscle lowers the leg medially, forming a U with the peroneus longus muscle (Angin et al., 2014).

The peroneus longus muscle, which crosses the foot, gives some stability to the sideways transverse and longitudinal arches of the foot. Its primary role is to evert the ankle, though it can also help with ankle plantar flexion. The shorter, smaller peroneus brevis muscle is located deep within the peroneus longus muscle. It connects sideways on the isolated fibula, lowers the leg, and circles behind the lateral malleolus before returning to the base of the fifth metatarsal. From the lateral malleolus onward, the peroneus brevis muscle is superficial (Botte et al., 1996). This muscle, like the peroneus longus, has the primary role of everting the ankle; however, it can also help with plantar flexion. The peroneus tertius muscle, which is not existent in everyone, is difficult to recognize and is sometimes mixed up with the extensor digitorum longus muscle. This muscle develops from the interosseous membrane and the distal medial fibula. It passes the ankle anteriorly to enclosure at the peroneus brevis muscle on the dorsal surface of the base of the 5th metatarsal. In theory, this muscle must evert and dorsiflex the ankle, however because of its small size, it is only helpful at best (Seale et al., 1988).

8.5.2. Intrinsic Muscles

Both attachments of intrinsic muscles are distal to the ankle joint. Because the muscles in the foot are not used to carry out complex activities, they are not as highly advanced as their equivalents in the hand. Their titles reveal a lot about their positions and activities. The extensor digitorum brevis, the dorsal interossei, and extensor hallucis brevis, which lie amongst the metatarsals and dorsal to the plantar interossei, are the only intrinsic muscles that are positioned on the plantar surface, virtually in layers (Attinger et al., 2002).

8.6. ANATOMICAL ASSOCIATIONS

To comprehend the interactions between the muscles of the ankle and foot, it is necessary to classify them into lateral, posterior, and anterior groups with deep and superficial subgroups. Six muscles organized in three layers comprise the posterior group. The sole superficial muscle that is positioned posteriorly is the gastrocnemius. It is surrounded by the extremely long, slender plantaris muscle and the massive, single-jointed soleus muscle. The flexor digitorum longus, flexor hallucis longus, and tibialis posterior are organized medially too laterally in the lowest stratum. As previously mentioned, the interdependence of these muscles alters twice more before to their insertions (Lawrence and Botte, 1994).

The peroneus longus is the most superficial of the adjacent group, whereas the peroneus brevis is the deepest. The peroneus brevis is located directly anterior to the peroneus longus, just above the lateral malleolus. Because it crosses the plantar surface of the foot deep, the peroneus longus cannot be observed or felt below the malleolus. The tendon of the peroneus brevis must come from behind the malleolus and the tendon of the peroneus tertius should come in front of the malleolus at the base of the fifth metatarsal. It is not to be confused with the extensor digitorum longus tendon. It should be noted that it does not extend to the fifth toe (Choplin et al., 2004).

The tibialis anterior muscle is an apparent muscle that instigates from the proximal lateral tibia and runs the whole length of the medial side of the ankle. The tendons of the extensor hallucis longus, extensor digitorum longus and tibialis anterior can be observed from medial to lateral just above the ankle. A tendon connects the 2nd, 3rd, 4th, and 5th toes to the extensor digitorum longus. Note the distinction between the extensor digitorum longus tendon that goes to the 5th toe and the peroneus tertius tendon that only goes to the base of the 5th metatarsal (Lawrence and Botte, 1995).

On the plantar surface, the foot's intrinsic muscles are organized in fundamentally four layers. The 1st layer of muscle lies beneath the plantar fascia. The flexor digitorum brevis is positioned in the middle of the foot, with tendons extending to the 2nd through 5th toes. The abductor hallucis is on the medial side, whereas the abductor digiti minimi is on the sideways. The second layer contains two extrinsic muscles and two intrinsic muscular tendons. The quadratus plantae extends from the calcaneus to the tendon of the flexor digitorum longus, where it joins right before the flexor digitorum longus divides into four tendons that travel to the 2nd through 5th toes (Boss and Hintermann, 2002).

The quadratus plantae adjusts the long toe flexor's pull line when it contracts. The flexor hallucis longus tendon is also visible in this stratum. The lumbricals are four intrinsic muscles that develop from the flexor digitorum longus tendons, travel through the medial side of the four lesser toes, and join on the dorsal surface to the tendons of the extensor digitorum longus (Milner and Soames, 1998). The flexor hallucis brevis has two heads medially, the adductor hallucis has two heads in the middle, and the flexor digiti minimi has two heads laterally. The interossei muscles make up the 4th and deepest layer. They are located between the bones on the dorsal and palmar sides, as their names suggest. They function similarly to their hand equivalents and have similar attachments. The second toe, unlike its equivalents in the hand, is the toe from which the rest toes abduct or adduct (Smith and Reischl, 1988).

The intrinsic muscles of the foot's dorsum are located beneath or close to their extrinsic counterparts. The extensor hallucis brevis is located directly across from the extensor hallucis longus. The three tendons of the extensor digitorum brevis attach laterally to the distal connection of the extensor digitorum longus on the 2nd, 3rd, and 4th toes (Figure 8.10) (Marsland et al., 2013).

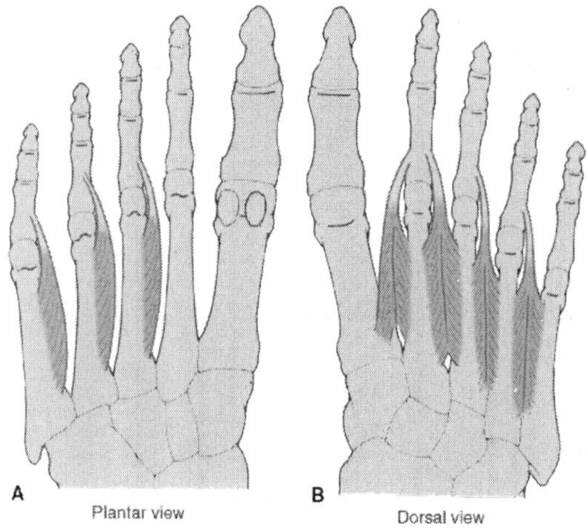

Figure 8.10. Muscles of the fourth (deepest) layer of the right foot's plantar surface. (a) Plantar interossei; (b) dorsal interossei.

Source: https://www.custompilatesandyoga.com/plantar-interossei/.

8.7. SUMMARY OF MUSCLE INNERVATION

According to innervation, the muscles of the ankle and foot are divided into rather neat groups. The tibial nerve supplies innervation to the muscles on the back of the leg and the plantar area of the foot. The plantar foot, like the hand, is divided into two groups. The lateral plantar division of the tibial nerve stimulates lateral side muscles, while the medial plantar branch innervates medial side muscles (Enoka, 1988).

Muscles on the lateral side of the leg are innervated by the superficial peroneal nerve. The peroneus tertius muscle is an exclusion, as it passes the ankle anteriorly and gains innervation from the deep peroneal nerve along with the other anterior muscles. As mentioned in previous chapters, there is some diversity in spinal cord level among sources. When a discrepancy occurs, Gray's Anatomy is employed as a reference source (Widmer et al., 2007).

8.8. USUAL ANKLE PATHOLOGIES

Shin splints are a broad name for pain caused by exercise alongside the medial border of the tibia, typically just a few inches above the ankle to halfway up the tibia. The most common cause of pain is periosteal inflammation. Shin splints are a repetitive stress injury caused by running on tiptoes, running on hard surfaces, and participating in activities that require a lot of hopping. A more precise term for anterior leg pain which is not caused by a stress fracture is medial tibial stress syndrome (Maquirriain, 2005).

Foot and toe deformities may influence the rest of the joints in the lower leg and trunk, particularly when walking or running. A plantigrade foot is one in which the sole is at 90 to the leg when the person is standing. Equinus foot refers to a permanent plantar flexion of the hindfoot. A dorsiflexed calcaneus foot is fixed. Pes cavus mentions to a high arch, whereas pes planus refers to the removal of the medial longitudinal arch. Pathological alterations in which the great toe generates a valgus deformity induce hallux valgus (Lui, 2007).

Hallux rigidus is a progressive condition characterized by pain and limited range of motion in the first MTP joint. All of the MTP joints are super extended in the following smaller toe deformities: The PIP is bent and the DIP is stretched in hammer toe. Mallet toe, on the other hand, has a flexed DIP joint and an expanded PIP joint. The PIP and DIP joints of the claw toe are both flexed (Rawool and Nazarian, 2000).

Metatarsalgia is a term used to describe discomfort in the metatarsal heads. The pain is frequently described as a bruise or as walking on pebbles. Increased exercise frequently makes the ache worse. Morton's neuroma is instigated by aberrant pressure on the plantar digital nerves, which often occurs between the 3rd and 4th metatarsals' web area. This pressure can cause numbness and pain in the toe area, which gets worse when you do things like run. Turf toe is triggered by the great toe being pushed hyperextended at the MTP joint. Football, baseball, and soccer players are all known to have it (DiGiovanni et al., 2000).

The ankle is the commonly damaged joint in the human body. Among competitive and recreational athletes, sprains of the ankle are likely the usual injury, and the adjacent ligament is the commonly damaged ligament. When the foot is the plantar-flexed, upturned position, lateral or inversion sprains can occur. One or more of the three portions of the lateral ligament might be strained or ripped. Ankle fractures frequently result from tripping over an unanticipated obstacle or falling from a height, and they typically involve ankle twisting. Most typically, the lateral malleolus is indulged. A bimalleolar fracture involves both malleoli, whereas a trimalleolar fracture involves the malleoli as well as the posterior tibial lip (Lee et al., 2011).

Plantar fasciitis is a frequent overuse ailment that causes heel pain. The plantar fascia keeps the medial longitudinal arch in place and works as a shock absorber while you walk. On the plantar surface, the discomfort is commonly felt where the fascia connects to the calcaneus. Achilles tendonitis, an infection of the gastrocnemius-soleus tendon, might be a sign of an Achilles tendon rupture. The person loses the capacity to plantar flex the ankle after a full rupture. Have someone lie susceptible to with their feet off the table's edge to see if the tendon is intact. Squeeze the gastrocnemius muscle's belly button. If the tendon is intact, there will be some plantar flexion, but if the tendon is ruptured, there will be no motion (Golanó et al., 2010).

The calcaneocuboid, talonavicular, and talocalcaneal joints are fused together in a triple arthrodesis. It gives the foot medial-lateral (ML) stability and decreases pain in the subtalar joint, but it takes away eversion and inversion at the ankle. Since the talotibial joint has not been affected, ankle dorsiflexion and plantar flexion continue (Ogilvie-Harris and Reed, 1994).

REFERENCES

1. Akdemir, G., (2010). Thoracic and lumbar intraforaminal ligaments. *Journal of Neurosurgery: Spine, 13*(3), 351–355.
2. Amiel, D., Frank, C., Harwood, F., Fronek, J., & Akeson, W., (1983). Tendons and ligaments: A morphological and biochemical comparison. *Journal of Orthopedic Research, 1*(3), 257–265.
3. Amis, A. A., Bull, A. M. J., Gupte, C. M., Hijazi, I., Race, A., & Robinson, J. R., (2003). Biomechanics of the PCL and related structures: Posterolateral, posteromedial and meniscofemoral ligaments. *Knee Surgery, Sports Traumatology, Arthroscopy, 11*(5), 271–281.
4. Amis, A. A., Gupte, C. M., Bull, A. M. J., & Edwards, A., (2006). Anatomy of the posterior cruciate ligament and the meniscofemoral ligaments. *Knee Surgery, Sports Traumatology, Arthroscopy, 14*(3), 257–263.
5. Angin, S., Crofts, G., Mickle, K. J., & Nester, C. J., (2014). Ultrasound evaluation of foot muscles and plantar fascia in pes planus. *Gait & Posture, 40*(1), 48–52.
6. Angin, S., Mickle, K. J., & Nester, C. J., (2018). Contributions of foot muscles and plantar fascia morphology to foot posture. *Gait & Posture, 61*, 238–242.
7. Astephen, J. L., Deluzio, K. J., Caldwell, G. E., & Dunbar, M. J., (2008). Biomechanical changes at the hip, knee, and ankle joints during gait are associated with knee osteoarthritis severity. *Journal of Orthopedic Research, 26*(3), 332–341.
8. Attinger, C. E., Ducic, I., Cooper, P., & Zelen, C. M., (2002). The role of intrinsic muscle flaps of the foot for bone coverage in foot and ankle defects in diabetic and nondiabetic patients. *Plastic and Reconstructive Surgery, 110*(4), 1047–1054.
9. Banerjee, K. K., & Agarwal, B. B. L., (1998). Estimation of age from epiphyseal union at the wrist and ankle joints in the capital city of India. *Forensic Science International, 98*(1, 2), 31–39.
10. Boss, A. P., & Hintermann, B., (2002). Anatomical study of the medial ankle ligament complex. *Foot & Ankle International, 23*(6), 547–553.
11. Botte, M. J., Santi, M. D., Prestianni, C. A., & Abrams, R. A., (1996). Ischemic contracture of the foot and ankle: Principles of management and prevention. *Orthopedics, 19*(3), 235–244.

12. Braunstein, B., Arampatzis, A., Eysel, P., & Brüggemann, G. P., (2010). Footwear affects the gearing at the ankle and knee joints during running. *Journal of Biomechanics, 43*(11), 2120–2125.
13. Cheung, J. T. M., & Zhang, M., (2006). Finite element modeling of the human foot and footwear. In: *ABAQUS Users' Conference* (Vol. 20, No. 6, pp. 145–158).
14. Choplin, R. H., Buckwalter, K. A., Rydberg, J., & Farber, J. M., (2004). CT with 3D rendering of the tendons of the foot and ankle: Technique, normal anatomy, and disease. *Radiographics, 24*(2), 343–356.
15. Chubinskaya, S., Kuettner, K. E., & Cole, A. A., (1999). Expression of matrix metalloproteinases in normal and damaged articular cartilage from human knee and ankle joints. *Laboratory Investigation; A Journal of Technical Methods and Pathology, 79*(12), 1669–1677.
16. Close, J. R., (1956). Some applications of the functional anatomy of the ankle joint. *JBJS, 38*(4), 761–781.
17. DeLancey, J. O., (1989). Pubovesical ligament: A separate structure from the urethral supports ("pubo-urethral ligaments"). *Neurourology and Urodynamics, 8*(1), 53–61.
18. DiGiovanni, B. F., Fraga, C. J., Cohen, B. E., & Shereff, M. J., (2000). Associated injuries found in chronic lateral ankle instability. *Foot & Ankle International, 21*(10), 809–815.
19. Donatelli, R., (1985). Normal biomechanics of the foot and ankle. *Journal of Orthopedic & Sports Physical Therapy, 7*(3), 91–95.
20. Eerdekens, M., Deschamps, K., & Staes, F., (2019). The impact of walking speed on the kinetic behavior of different foot joints. *Gait & Posture, 68*, 375–381.
21. Enoka, R. M., (1988). Muscle strength and its development. *Sports Medicine, 6*(3), 146–168.
22. Floyd, R. T., & Thompson, C. W., (2009). *Manual of Structural Kinesiology* (Vol. 16, pp. 1–7). New York, NY: McGraw-Hill.
23. Fong, D. T. P., Chan, Y. Y., Hong, Y., Yung, P. S. H., Fung, K. Y., & Chan, K. M., (2008). A three-pressure-sensor (3PS) system for monitoring ankle supination torque during sport motions. *Journal of Biomechanics, 41*(11), 2562–2566.
24. Gabbard, C., & Hart, S., (1996). A question of foot dominance. *The Journal of General Psychology, 123*(4), 289–296.
25. Gheitasi, M., Maleki, M., & Bayattork, M., (2022). Corrective exercise

for intrinsic foot muscles versus the extrinsic muscles to rehabilitate flat foot curving in adolescents: Randomized-controlled trial. *Sport Sciences for Health*, 1–10.

26. Glassow, R. B., (1966). Kinesiology: Modern dance and Kinesiology. *Journal of Health, Physical Education, Recreation, 37*(1), 65–68.
27. Golanó, P., Vega, J., De Leeuw, P. A., Malagelada, F., Manzanares, M. C., Götzens, V., & Van, D. C. N., (2010). Anatomy of the ankle ligaments: A pictorial essay. *Knee Surgery, Sports Traumatology, Arthroscopy, 18*(5), 557–569.
28. Guillén-Rogel, P., San, E. C., & Marín, P. J., (2017). Associations between ankle dorsiflexion range of motion and foot and ankle strength in young adults. *Journal of Physical Therapy Science, 29*(8), 1363–1367.
29. Gupte, C. M., Bull, A. M., Thomas, R. D., & Amis, A. A., (2003). A review of the function and biomechanics of the meniscofemoral ligaments. *Arthroscopy: The Journal of Arthroscopic & Related Surgery, 19*(2), 161–171.
30. Hargrove, L. J., Simon, A. M., Lipschutz, R. D., Finucane, S. B., & Kuiken, T. A., (2011). Real-time myoelectric control of knee and ankle motions for transfemoral amputees. *JAMA, 305*(15), 1542–1544.
31. Hendren, L., & Beeson, P., (2009). A review of the differences between normal and osteoarthritis articular cartilage in human knee and ankle joints. *The Foot, 19*(3), 171–176.
32. Hunt, A. E., Smith, R. M., & Torode, M., (2001). Extrinsic muscle activity, foot motion and ankle joint moments during the stance phase of walking. *Foot & Ankle International, 22*(1), 31–41.
33. Huson, A., (1987). Joints and movements of the foot: Terminology and concepts. *Acta Morphol. Neerl. Scand., 25*(3), 117–130.
34. Jackson, K., Simon, J. E., & Docherty, C. L., (2016). Extended use of Kinesiology tape and balance in participants with chronic ankle instability. *Journal of Athletic Training, 51*(1), 16–21.
35. Jones, O. M., Smeulders, N., Wiseman, O., & Miller, R., (1999). Lateral ligaments of the rectum: An anatomical study. *British Journal of Surgery, 86*(4), 487–489.
36. Kim, B. J., Lee, J. H., Kim, C. T., & Lee, S. M., (2015). Effects of ankle balance taping with Kinesiology tape for a patient with chronic ankle instability. *Journal of Physical Therapy Science, 27*(7), 2405, 2406.

37. Kim, M. K., & Shin, Y. J., (2017). Immediate effects of ankle balance taping with Kinesiology tape for amateur soccer players with lateral ankle sprain: A randomized cross-over design. *Medical Science Monitor: International Medical Journal of Experimental and Clinical Research, 23*, 5534.

38. Kipp, K., & Palmieri-Smith, R. M., (2013). Differences in kinematic control of ankle joint motions in people with chronic ankle instability. *Clinical Biomechanics, 28*(5), 562–567.

39. Kirby, K. A., (2017). Longitudinal arch load-sharing system of the foot. *Revista Española de Podología, 28*(1), e18–e26.

40. Klaue, K., (1997). Planovalgus and cavovarus deformity of the hind foot. *The Journal of Bone and Joint Surgery; British, 79*(6), 892–895.

41. Kurihara, T., Yamauchi, J., Otsuka, M., Tottori, N., Hashimoto, T., & Isaka, T., (2014). Maximum toe flexor muscle strength and quantitative analysis of human plantar intrinsic and extrinsic muscles by a magnetic resonance imaging technique. *Journal of Foot and Ankle Research, 7*(1), 1–6.

42. Kwon, S., & Park, J., (2012). Kinesiology-based robot foot design for human-like walking. *International Journal of Advanced Robotic Systems, 9*(6), 259.

43. Lawrence, S. J., & Botte, M. J., (1994). The sural nerve in the foot and ankle: An anatomic study with clinical and surgical implications. *Foot & Ankle International, 15*(9), 490–494.

44. Lawrence, S. J., & Botte, M. J., (1995). The deep peroneal nerve in the foot and ankle: An anatomic study. *Foot & Ankle International, 16*(11), 724–728.

45. Lee, B. G., & Lee, J. H., (2015). Immediate effects of ankle balance taping with Kinesiology tape on the dynamic balance of young players with functional ankle instability. *Technology and Health Care, 23*(3), 333–341.

46. Lee, J., Hamilton, G., & Ford, L., (2011). Associated intra-articular ankle pathologies in patients with chronic lateral ankle instability: Arthroscopic findings at the time of lateral ankle reconstruction. *Foot & Ankle Specialist, 4*(5), 284–289.

47. Long, Z., Wang, R., Han, J., Waddington, G., Adams, R., & Anson, J., (2017). Optimizing ankle performance when taped: Effects of

Kinesiology and athletic taping on proprioception in full weight-bearing stance. *Journal of Science and Medicine in Sport, 20*(3), 236–240.

48. Loudon, J. K., & Bell, S. L., (1996). The foot and ankle: An overview of arthrokinematics and selected joint techniques. *Journal of Athletic Training, 31*(2), 173.
49. Lui, T. H., (2007). Arthroscopy and endoscopy of the foot and ankle: Indications for new techniques. *Arthroscopy: The Journal of Arthroscopic & Related Surgery, 23*(8), 889–902.
50. Maquirriain, J., (2005). Posterior ankle impingement syndrome. *JAAOS-Journal of the American Academy of Orthopedic Surgeons, 13*(6), 365–371.
51. Marsland, D., Dray, A., Little, N. J., & Solan, M. C., (2013). The saphenous nerve in foot and ankle surgery: Its variable anatomy and relevance. *Foot and Ankle Surgery, 19*(2), 76–79.
52. McKeon, P. O., Hertel, J., Bramble, D., & Davis, I., (2015). The foot core system: A new paradigm for understanding intrinsic foot muscle function. *British Journal of Sports Medicine, 49*(5), 290–290.
53. Milner, C. E., & Soames, R. W., (1998). Anatomy of the collateral ligaments of the human ankle joint. *Foot & Ankle International, 19*(11), 757–760.
54. Nogueira, M. P., Farcetta, F., & Zuccon, A., (2015). Cavus foot. *Foot and Ankle Clinics, 20*(4), 645–656.
55. Oatis, C. A., (1988). Biomechanics of the foot and ankle under static conditions. *Physical Therapy, 68*(12), 1815–1821.
56. Ogilvie-Harris, D. J., & Reed, S. C., (1994). Disruption of the ankle syndesmosis: Diagnosis and treatment by arthroscopic surgery. *Arthroscopy: The Journal of Arthroscopic & Related Surgery, 10*(5), 561–568.
57. Park, D. J., & Park, S. Y., (2018). Comparison of subjects with and without pes planus during short foot exercises by measuring muscular activities of ankle and navicular drop height. *Journal of the Korean Society of Physical Medicine, 13*(3), 133–139.
58. Poynton, A. R., Javadpour, S. M., Finegan, P. J., & O'brien, M., (1997). The meniscofemoral ligaments of the knee. *The Journal of Bone and Joint Surgery; British, 79*(2), 327–330.

59. Puno, R. M., Vaughan, J. J., Stetten, M. L., & Johnson, J. R., (1991). Long-term effects of tibial angular malunion on the knee and ankle joints. *Journal of Orthopedic trauma, 5*(3), 247–254.
60. Rawool, N. M., & Nazarian, L. N., (2000). Ultrasound of the ankle and foot. In: *Seminars in Ultrasound, CT and MRI* (Vol. 21, No. 3, pp. 275–284). WB Saunders.
61. Roaas, A., & Andersson, G. B., (1982). Normal range of motion of the hip, knee and ankle joints in male subjects, 30–40 years of age. *Acta Orthopedica Scandinavica, 53*(2), 205–208.
62. Rougier, P., & Caron, O., (2000). Center of gravity motions and ankle joint stiffness control in upright undisturbed stance modeled through a fractional Brownian motion framework. *Journal of Motor Behavior, 32*(4), 405–413.
63. Rumian, A. P., Wallace, A. L., & Birch, H. L., (2007). Tendons and ligaments are anatomically distinct but overlap in molecular and morphological features—A comparative study in an ovine model. *Journal of Orthopedic Research, 25*(4), 458–464.
64. Sarrafian, S. K., (1987). Functional characteristics of the foot and plantar aponeurosis under tibiotalar loading. *Foot & Ankle, 8*(1), 4–18.
65. Scarton, A., Guiotto, A., Malaquias, T., Spolaor, F., Sinigaglia, G., Cobelli, C., & Sawacha, Z., (2018). A methodological framework for detecting ulcers' risk in diabetic foot subjects by combining gait analysis, a new musculoskeletal foot model and a foot finite element model. *Gait & Posture, 60*, 279–285.
66. Schweitzer, M. E., Van, L. M., Ehrlich, S. S., & Wapner, K., (1994). Fluid in normal and abnormal ankle joints: Amount and distribution as seen on MR images. *American Journal of Roentgenology (AJR), 162*(1), 111–114.
67. Seale, K. S., Lange, T. A., Monson, D., & Hackbarth, D. A., (1988). Soft tissue tumors of the foot and ankle. *Foot & Ankle, 9*(1), 19–27.
68. Sereno, P. C., & Arcucci, A. B., (1990). The monophyly of crurotarsal archosaurs and the origin of bird and crocodile ankle joints. *Neues Jahrbuch für Geologie und Paläontologie, Abhandlungen, 180*(1), 21–52.
69. Smith, R. W., & Reischl, S., (1988). The influence of dorsiflexion in the treatment of severe ankle sprains: An anatomical study. *Foot & Ankle, 9*(1), 28–33.

70. Soysa, A., Hiller, C., Refshauge, K., & Burns, J., (2012). Importance and challenges of measuring intrinsic foot muscle strength. *Journal of Foot and Ankle Research, 5*(1), 1–14.
71. Widmer, C. G., English, A. W., & Morris-Wiman, J., (2007). Developmental and functional considerations of masseter muscle partitioning. *Archives of Oral Biology, 52*(4), 305–308.
72. Woo, S. L., Debski, R. E., Withrow, J. D., & Janaushek, M. A., (1999). Biomechanics of knee ligaments. *The American Journal of Sports Medicine, 27*(4), 533–543.
73. Zhang, L. Q., Chung, S. G., Bai, Z., Xu, D., Van, R. E. M., Rogers, M. W., & Roth, E. J., (2002). Intelligent stretching of ankle joints with contracture/spasticity. *IEEE Transactions on Neural Systems and Rehabilitation Engineering, 10*(3), 149–157.

INDEX

A

Abduction 107, 108, 109, 113, 115, 116
abductor digiti minimi 115, 116
acromioclavicular joint 36, 53, 54, 55, 57, 69, 71, 73
adduction 107, 108, 109, 113, 116
anatomical position 107
anatomy 2, 5, 24
ankle 122, 123, 124, 126, 127, 128, 129, 132, 135, 137, 138, 146, 147
Anterior 206
Anterior Inferior Iliac Spine (AIIS) 175
anterior-posterior (AP) 36, 150
anterior scalene muscle 159
anterior superior iliac spine (ASIS) 174
anterior talofibular ligament 204
aperture 154
auxiliary motion 107

B

Biomechanics 2, 4, 28, 29, 30

C

capitulum 77, 78, 79, 82, 86, 97, 99, 100, 103
carpometacarpal (CMC) 106, 108
Cervical protraction 152
Chinese acupuncture 5
clavicle 34, 35, 36, 38, 39, 45, 46, 48, 49, 50, 51, 52, 53, 54, 55, 57, 58, 59, 60, 68, 72, 74
coastal facets 155
coccyx 172, 173, 182
coronoid bone 80
costal surface 37
curiosity 4

D

Deep posterior 206
deltoid tuberosity 77
distal humerus 76, 77, 78, 79, 80, 86, 99, 102
distal interphalangeal (DIP) 107
distal ulna 80, 101
Dorsiflexion 122
dynamic systems 2

E

elbow 76, 77, 78, 80, 82, 83, 84, 85, 86, 87, 88, 89, 90, 91, 92, 93, 94, 95, 96, 97, 98, 99, 100, 101, 102, 103, 104
eversion 198, 199, 201, 202, 209, 214
extensor retinaculum ligament 110
Extinction 173
extrinsic muscles 111, 114, 118, 120

F

femur 172, 174, 176, 178, 179, 180, 181, 182, 183, 186
fibrocartilaginous acetabular labrum 179, 180
Fibrous tissue 199
fibula 194, 195, 199, 200, 207, 208, 209
fingers 106, 107, 108, 109, 110, 112, 113, 114, 115, 116, 118, 120
flexion 107, 108, 109, 113, 115, 118
flexor digitorum profundus muscle 113
foot 122, 123, 124, 125, 126, 127, 129, 130, 132, 133, 135, 137, 138, 139, 140, 141, 145, 147
Foot flat 126, 127
forearm 107, 108, 113
fused bones 172

G

gait cycle 123, 124, 125, 126, 130, 133, 140, 144, 145, 146
geometry 2
glenoid fossa 40, 41, 42, 56
gravity 130, 133, 134, 135

H

Heel-off 126
Heel striking 126
horizontal displacement 130, 132, 139
humerus 34, 38, 42, 43, 44, 46, 54, 61, 67
humerus flattening 77
Hyaline cartilage 79
hyperextension 173, 178, 182

I

ilium 172, 174, 182, 183
infrahyoid muscles 164
infraspinous fossa 38
interphalangeal (IP) joints 106
interspinal ligament 158
Intrinsic muscles 114
Inversion 198, 201
ischium 172, 174, 175, 176, 182

J

joints 34, 36, 45, 47, 52, 54, 55, 57, 58, 64, 66, 67, 73

K

keystone 205
Kinematics 2
Kinesiology 1, 2, 3, 4, 5, 6, 7, 9, 23, 24, 25, 26, 27, 28, 29, 30, 31
Kinesiology employs mechanics 2
Kinetics 2, 30

L

Lateral groups/compartments 206
lateral ligament 204, 213
Leg advancement 125
ligamentum teres 179, 192
longissimus muscle group 163

long planar ligament 206
lymphatic drainage 5

M

metacarpophalangeal (MCP) 106
Mid-stance 126
Midswing 129
mobility 34, 35, 45, 47, 50, 55, 56, 64, 197
mortise 199
multijointed pole 150
muscles 34, 35, 36, 39, 40, 43, 46, 47, 54, 56, 57, 58, 61, 62, 63, 64, 65, 66
musculoskeletal anatomy 2

N

neurology 5
neuromuscular physiology 2
nutrition 5

O

observation 4
olecranon 77, 78, 80, 81, 85
olecranon fossa 77, 78, 85
organs 83
os coxae 172, 174

P

passion 4
patella 172, 177, 181
Pathology 132
pelvis bones 172
pelvis girdle 173
physics 2, 13, 24
physiology 2, 5, 24
physiotherapy 4
plantar flexion 122, 123, 126, 127, 128, 137
Plantar flexion 198, 200
platysma muscle 164
Posterior Inferior Iliac Spine (PIIS) 175
Posterior Superior Iliac Spine (PSIS) 175
posterior surface 36, 37, 38
posterior talofibular ligament 204
prevertebral muscles 151, 159, 164
Pronation 123
proximal interphalangeal (PIP) 107
proximal radius 76, 77, 81, 82
proximal ulna 76, 77, 80, 85
pubis 172, 174, 175, 176, 178, 181, 182

Q

quadratus lumborum muscle 158, 164

R

radial groove 77
Rancho Los Amigos (RLA) 125
real flexion 122
rectus abdominis muscle 161
robust interosseous membrane 194

S

sacrum 172, 173, 182
scapula 34, 35, 36, 37, 38, 39, 40, 41, 42, 45, 46, 47, 53, 54, 55, 56, 57, 58, 59, 60, 61, 62, 63, 64, 67, 73
Shin splints 212
shoulder complex 34, 35, 42, 43, 44, 46, 47, 48, 52, 53, 55, 58, 60, 63, 64, 66, 68, 71
shoulder dysfunction 34

shoulders 131, 135
skull 150, 151, 152, 154, 160, 169
spinal column 150, 151, 152, 156, 159, 160, 163, 164
spinal cord 150, 153
Splenius capitis 160
splenius cervicis muscles 160
sternocleidomastoid muscle 152, 158, 159, 164
stride 123, 132, 135, 137, 139
suboccipital muscles 152, 160
Superficial posterior 206
superior-inferior (SI) 36
supination 82, 84, 88, 90, 92, 94, 95, 96, 104
supraspinous fossa 38

T

thorax 34, 35, 36, 37, 39, 44, 45, 46, 47, 56, 57, 63
thorax morphology 34
Thumb joints 106
Thumb movements 107
tibia 194, 195, 199, 200, 201, 207, 208, 209, 210, 212
Toe-off 126, 128
triaxial joint 173
Triplanar motion 200
trochlea 77, 78, 79, 80, 81, 84, 85, 86, 97
trochlear notch 80, 81, 84, 85, 97, 98
true ankle joint 199
trunk 122, 126, 127, 131, 133, 135, 136, 139

U

ulnar crest 80
ulnar nerve 78, 103

V

vertebral column 150, 154, 156, 158, 164
vertical axis 151
Vertical displacement 132

W

Wolff's law 79